"The average cognitive scientist today may not be familiar with Naomi Weisstein, yet the influence of her brief career in science has been widespread and profound. The contributors to this volume — an all-star team of modern vision scientists — trace many of the guiding themes in today's research back to this source. An important clue is that the scientist known as Naomi was also a comedian, a pioneering feminist, and a health activist. These chapters will help today's scientists understand how such a unique personality transformed the study of the human mind."

Professor James T. Enns,
University of British Columbia, Canada

PIONEER VISUAL NEUROSCIENCE

This book honors Naomi Weisstein's foreshortened span of work published from 1964 to 1992. Naomi Weisstein was a pioneer in the areas we now call visual neuroscience, visual cognition, and cognitive neuroscience. Her enthusiastic pursuit of the mind was infectious, inspiring many others to take up the challenge. Despite her time as an active researcher being cut short, Weisstein's impact was far reaching and long lasting, and many of her ideas and insights foreshadowed today's active areas of inquiry into the inner workings of the mind.

Comprising contributions from leading scholars in the field, Pioneer Visual Neuroscience outlines Weisstein's many contributions to the study of visual perception and processing and their effects on the field today.

This volume will be of interest to anyone interested in visual perception, visual cognition, and cognitive neuroscience.

James M. Brown is Professor in the Department of Psychology at the University of Georgia. He received his PhD from the University at Buffalo where he was Naomi Weisstein's last graduate student.

PIONEER VISUAL NEUROSCIENCE

A Festschrift for Naomi Weisstein

Edited by James M. Brown

Routledge
Taylor & Francis Group
NEW YORK AND LONDON

First published 2019
by Routledge
711 Third Avenue, New York, NY 10017

and by Routledge
2 Park Square, Milton Park, Abingdon, Oxon, OX14 4RN

Routledge is an imprint of the Taylor & Francis Group, an informa business

Library of Congress Cataloging-in-Publication Data
A catalog record for this title has been requested

ISBN: 978-1-138-04838-6 (hbk)
ISBN: 978-1-138-04839-3 (pbk)
ISBN: 978-1-315-17018-3 (ebk)

Typeset in Bembo
by Apex CoVantage, LLC

Cover: the cover is based on two of Naomi Weisstein's Science papers, "Neural Symbolic Activity" (1970) and "The Object-Superiority Effect" (1974) and was designed with the assistance of Joanna Eberts, http://www.joanna-eberts.com/ and utilizes an original picture of Weisstein taken in 1972 by Virginia Blasedell, http://www.vblaisdell.com/.

CONTENTS

LIST OF MOVIES

The following movies can be accessed through:

https://research.franklin.uga.edu/visual-perception-laboratory/festschrift-naomi-weisstein

For Chapter 2:

Original Phantoms Xs
Original Light Dark Gratings Phantoms
Phantoms in the Lab

For Chapter 4:

Movie 1
Movie 2

For Chapter 5:

Video 1
Video 2A
Video 2B
Video 2C
Video 3
Video 4
Video 5
Video 6
Video 7A
Video 7B
Video 7C
Video 8
Video 9A

INTRODUCTION

*James M. Brown and Patrick Cavanagh**
*University of Georgia, *Glendon College*

Let's start with some of the nitty gritty details of Naomi Weisstein's career before filling out this skeleton with the impact of her work and the substance of what she accomplished. Naomi received her B.A. with special honors from Wellesley College in 1961 graduating Phi Beta Kappa. As a doctoral student at Harvard University, she won a Departmental Distinction award (1961–1962), was an Honorary Woodrow Wilson Fellow, and was a National Science Foundation Predoctoral Fellow from 1961 until graduating at the top of her class with her Ph.D. in 1964. Earning a Ph.D. in two and a half years from any institution, let alone Harvard, is remarkable for anyone past or present! In the fall of 1964 Naomi was a United States Public Health Services Postdoctoral Trainee in Mathematical Biology at the University of Chicago, becoming a National Science Foundation Postdoctoral Fellow with the Committee on Mathematical Biology at the University of Chicago from January to October of 1965. From 1965–1966 she was a Lecturer and Research Associate with the Collegiate Division of the Social Sciences and a research Associate with the Department of Psychiatry at the University of Chicago. From 1966 to 1973 she was a professor in the Department of Psychology at Loyola University, Chicago. From 1973 until her retirement due to illness, Weisstein was a Professor in the Department of Psychology's cognitive program at the State University of New York at Buffalo and became Professor Emeritus in 2003. While at the University at Buffalo she was awarded a Guggenheim Fellowship (1980–1981) and was elected a Fellow of the American Association for the Advancement of Science (1981) and a Fellow of the American Psychological Association (1995). Also, during this time, she was awarded grants from the National Science Foundation (NSF), National Institutes of Health and the National Eye Institute (NEI), and Defense Advanced Research Projects Agency (DARPA).

Naomi had a remarkable impact on her field and on the conduct of science in general. In the 1970s, she was an emerging leader in the relatively new fields of visual cognition and cognitive neuroscience. She was passionate about science. She loved scientific research not only because it presented puzzles that she could crack open – at times faster and better than the rest of us – but because if you discovered something new and interesting, it would be recognized, no matter who you were or where you were from, and especially if you were challenging the theories of some of the more powerful players in your field. Of course, that was the ideal that she found so attractive; but at that time the truth was far from the ideal. The sexism she encountered in the scientific community became one of the targets of her feminist crusades. Many fell in her take-no-prisoner assaults against sexism. Deans and advisors, colleagues who used slides of scantily clad women and other sexist jokes as traditional material in talks and lectures found their presentations derailed by the Weisstein buzz-saw of gender justice (e.g., see Weisstein, 1977). Naomi made a difference and she made it in many ways, on many levels.

She was one of the two or three most influential people in cognitive science. Her presentations were not only brilliant for their content but they were also drop-dead funny, as all who knew her can imagine. Humor is not usually part of scientific presentations but it always was for Naomi. She brought so many new recruits into our area, us included, by making us think, then laugh. We thought any area led by someone with so much charisma must be really the place to be.

Part of her success was getting many articles published in top level journals like *Science*, one of the most exclusive journals in the world. Anyone with only a single article in *Science* in their career is considered a real success. So here are some data, the kind of real evidence that Naomi loved so much. Our colleague Neil deGrasse Tyson has had one article in *Science*, Jane Goodall has had four. What about Nobel Laureates? David Hubel had two, Daniel Kahneman had six, Francis Crick had four, Murray Gell-Mann had one. Naomi? She had six papers in *Science* in her brief career, a truly outstanding achievement.

What were they about? They were about the mind, how it works, how it creates the visual world we see. About a third of our brain is dedicated to seeing and that powerful computational machinery is what makes seeing seem so effortless to us. Understanding that machinery was Naomi's pursuit. Across several projects she showed that the representation in the brain was more holistic and complete than what is presented in the scene in front of us. Hidden details are fully represented. We might expect that, but Naomi showed it was true by revealing these behind-the-scenes completion effects. For example, visual details were better recognized as parts of objects than in isolation or as parts of non-objects. Objectness was a property of things that allowed recognizing things like words not just from their letters but also from the their global shapes, and from our knowledge of words (Weisstein & Harris, 1974; Weisstein et al., 1982; A. Williams & Weisstein, 1978; M. C. Williams & Weisstein, 1984, see Chapter 3 (Breitmeyer) related to objects and Chapters 8 (Wolfe) and 9 (Fines & Reeves), related to objects and attention).

Patterns that were hidden from view in one region of a display would still mask targets presented in that region, making them hard to see (Weisstein, 1970a, 1970b; Weisstein et al., 1972, see Chapter 6 (Ogmen) for more on masking and Chapter 7 (Wolfson & Graham) for more on "hidden" intermediate processes). She introduced us to the power of small dots that would interfere with the visibility of other distant targets, these were her "killer masks" that supported the idea that visual stimuli were made up of (invisible) waves – a tiny dot can be shown to be composed of a broad spectrum of waves (i.e., sinewaves of varying spatial frequency, amplitude, orientation, and phase) (Weisstein & Bisaha, 1972; Weisstein & Harris, 1980; Weisstein et al., 1977). She also introduced us to our ability to see things that were not present but should be, like in the motion-induced contour where a clear contour is seen even though none is present (Klymenko & Weisstein, 1981, see Chapter 5 (Erlikhman & Caplovitz)) and, in "phantom gratings" where vertical bars were interrupted in the middle by a gap but when the visible parts started moving, the bars would visibly fill in across the gap. In one of her most dramatic presentations, she added Xs to the visible parts above and below the gap, and when these bars composed of Xs started to move, observers saw not just the bars completed in the empty region, but also actual Xs, as if hallucinating fully formed letters that were not there at all (Weisstein & Harris, 1980; Weisstein et al., 1974, and see Chapters 3 (Breitmeyer), 2 (Cavanagh), and 4 (Gyoba, Sakurai, & Kitaoka)). Truly, science can be more magical than magic. A final example of the legacy of Naomi's thinking is the dynamic, interactive nature of figure-ground perception revealed by how the spatial and temporal nature of sensory stimulation biased our perception of a region towards figure (e.g., an object) or ground (Klymenko & Weisstein, 1986; Wong & Weisstein, 1987) and how perceiving a region as figure or ground altered our sensory response to those regions (Brown & Weisstein, 1988; Wong & Weisstein, 1982, 1983). (See Chapters 1 (Brown & Greene), 10 (Peterson), and 11 (Gillam) for more on bottom-up/top-down interactions in figure-ground perception.)

In the early 1980s Naomi began to suffer from what was eventually determined to be Chronic Fatigue Immune Dysfunction (CFIDS) (also known as Chronic Fatigue Syndrome (CFS) and Myalgic Encephalomyelitis (ME)). The beginning of her illness led to lab meetings in her living room while she laid on the couch, to her return to NYC with her husband Jesse Lemisch, and ultimately to her being bedridden for the next thirty plus years until her death in the spring of 2015. One wonders where the field of vision science would be today had she remained healthy and been able to spend those years enthusiastically pursuing her study of the mind.

A big challenge organizing this Festschrift was doing justice to Naomi because of the many ways she influenced and inspired so many people as a leader in both vision science and the women's movement. She also inspired us by her courage battling CFIDS for so long all the while keeping her wit, sense of humor, and interest in the two frontiers she blazed that were integral to who she was.

While this Festschrift focuses on Naomi's influence, inspiration, and accomplishments in science, it is important to be mindful of the context within which her accomplishments were made (i.e., sexism, bigotry, misogyny). To gain greater appreciation of that context and how truly amazing she was, the reader is strongly encouraged to explore the links at: https://research.franklin.uga.edu/visual-perception-laboratory/introduction

A Festschrift honors an individual's lifetime of work oftentimes spanning four or five decades or more. Our Festschrift (or more accurately, our Gedenkschrift, i.e., after the honoree has deceased) honors Weisstein's foreshortened span of published work from 1964 to 1992. Despite her time as an active researcher being cut short, as the following chapters attest, Weisstein's impact was far reaching and long lasting. In many ways her ideas and insights foreshadowed many of today's active areas of inquiry into the inner workings of the mind.

References

Brown, J. M. & Weisstein, N. (1988). A phantom context effect: Visual phantoms enhance target visibility. *Perception & Psychophysics*, *43*(1), 53–56.

Klymenko, V. & Weisstein, N. (1981). The motion-induced contour. *Perception*, *10*(6), 627–636.

Klymenko, V. & Weisstein, N. (1986). Spatial frequency differences can determine figure-ground organization. *Journal of Experimental Psychology Human Perception & Performance*, *12*(3), 324–330.

Weisstein, N. (1970a). Neural symbolic activity: A psychophysical measure. *Science*, *168*(3938), 1489.

Weisstein, N. (1970b). Reply to Sekuler and Armstrong. *Science*, *170*, 1227–1228.

Weisstein, N. (1977). "How can a little girl like you teach a great big class of men?" the chairman said, and the other adventures of a woman in science. In S. Ruddick & P. Daniels (Eds), *Working it out: 23 writers, artists, scientists, and scholars talk about their lives and work* (pp. 241–250). New York: Pantheon Books.

Weisstein, N. & Bisaha, J. (1972). Gratings mask bars and bars mask gratings: Visual frequency response to aperiodic stimuli. *Science*, *176*(4038), 1047–1049.

Weisstein, N. & Harris, C. S. (1974). Visual detection of line segments: An object-superiority effect. *Science*, *186*(4165), 752.

Weisstein, N. & Harris, C. S. (1980). Masking and unmasking of distributed representations in the visual system. In C. S. Harris (Ed.), *Visual coding and adaptability* (pp. 317–364). Mahwah, NJ: Lawrence Erlbaum Associates.

Weisstein, N., Mathews, M. & Berbaum, K. (1974). Illusory contours can mask real contours. *Bulletin of the Psychonomic Society*, *4*, 266.

Weisstein, N., Montalvo, F. & Ozog, G. (1972). Differential adaptation to gratings blocked by cubes and gratings blocked by hexagons: A test of the neural symbolic activity hypothesis. *Psychonomic Science*, *27*(2), 89–91.

Weisstein, N., Williams, M. C. & Harris, C. S. (1982). Depth, connectedness, and structural relevance in the object-superiority effect: Line segments are harder to see in flatter patterns. *Perception*, *11*(1), 5–17.

Weisstein, N., Harris, C. S., Berbaum, K., Tangney, J. & Williams, A. (1977). Contrast reduction by small localized stimuli: Extensive spatial spread of above-threshold orientation-selective masking. *Vision Research*, *17*(3), 341–350.

Williams, A. & Weisstein, N. (1978). Line segments are perceived better in a coherent context than alone: An object-line effect in visual perception. *Memory & Cognition*, *6*(2), 85–90.

Williams, M. C. & Weisstein, N. (1984). The effect of perceived depth and connectedness on metacontrast functions. *Vision Research*, *24*(10), 1279–1288.

Wong, E. & Weisstein, N. (1982). A new perceptual context-superiority effect: Line segments are more visible against a figure than against a ground. *Science*, *218*(4572), 587.

Wong, E. & Weisstein, N. (1983). Sharp targets are detected better against a figure, and blurred targets are detected better against a background. *Journal of Experimental Psychology: Human Perception & Performance*, *9*(2), 194–201.

Wong, E. & Weisstein, N. (1987). The effects of flicker on the perception of figure and ground. *Perception & Psychophysics*, *41*(5), 440–448.

1

WE'RE GOING TO STUDY THE MIND!

*James M. Brown and Harold H. Greene**
*University of Georgia, *University of Detroit Mercy*

Introduction

It all started for me (James) the spring of 1978, when, on the first day of my last required course for my undergraduate degree in psychology, Professor Naomi Weisstein introduced herself to her Sensation and Perception class and said "We're going to study the mind!" "Really," I said to myself, "now that sounds interesting". Over the following weeks I realized I was hooked. This was very cool stuff, real psychological science. I needed to know more. I asked Dr. Weisstein after class how I might get more involved in this mind stuff. She directed me to her lab where her graduate students trained me to be an observer in their ongoing experiments. As it turned out the experiments Amanda and Mary Williams (no relation) trained me to run in were metacontrast masking studies exploring the temporal and configural nature of the object superiority effect. In the beginning I was truly a naïve observer in many ways. I wondered how what I was doing could possibly be of any help to them because I often felt I was simply guessing which of four target lines embedded in a figure I saw on each trial. Later during my first lab meeting, Naomi seemed thrilled with my results as much as I was amazed by them looking so systematic and similar to the other observer's results. This was my first experience with and introduction to what Naomi later described to me as cognitive psychophysics. Towards the end of that semester I asked Naomi where I might go from here given how fascinated I had become with the study of visual perception. Without a moment's hesitation she suggested I enroll as a graduate student and work with her at SUNY Buffalo. I was shocked and then so excited for the opportunity. From that first day of class and since, I have felt ever so lucky to have met this wonderful woman. As I have described to her beloved husband Jesse Lemisch, I was blessed to have crossed paths with this amazing woman, this ball of energy that was the "comet Naomi". Somehow space and time aligned for me to have crossed the tail of the comet Naomi as it streaked across the visual-cognition sky and in doing so my life was changed forever. As we know though, when we see a comet, it is in the process of burning out. That's why I think of myself as passing through the tail of the "comet Naomi" because soon after starting to work with her she began to suffer from what was later determined to be Chronic Fatigue Immune Dysfunction (CFIDS). Remarkably, as the comet Naomi was slowly physically burning out over so many years, her amazing heart and spirit continued to burn brightly. The beginning of her illness led to lab meetings in her living room while she laid on the couch, to her return to NYC with Jesse, and ultimately to her being bedridden for the next thirty plus years until her death in the spring of 2015. I often wonder where our field of vision science would be today had she remained healthy and been able to spend those years pursuing her study of the mind. It was her energy and enthusiasm for studying the mind that inspired me to study visual perception.

It is the seeming immediacy and effortlessness of visual perception that belies the amazing complexity of what underlies it. As we interact with our environment the stimulation received through our eyes informs us of the world around us. How this information is acquired and utilized by our brains resulting in our conscious experience is still a curious question. It certainly involves more than "simply" taking in stimulation from the environment. One might consider our visual brains as "reaching out" to engage with the world to be able to understand what is out there, analogous to reaching out with our hands to identify through touch something we are unable to see. In both instances the afferent stimulation via the senses is utilized within the context of directed brain activity to "grasp" what we believe to be the nature of the world we perceive based on expectations and experiences. The expectations and experiences are a mix of variable short-term ones that precede and guide our sequential fixations as we explore the visual environment and stable long-term ones based on our evolutionary past and our knowledge of the world accumulated via lifetime experiences of acting upon and moving about in the world (e.g., Hoffman, 1996; O'Regan & Noë, 2001; Purves et al., 2015). Trying to understand how this complex interplay between bottom-up stimulation and top-down experience and knowledge creates our perceptual experience underlies most of Naomi Weisstein's research and illustrates her influence on our (Naomi's academic genealogical son and grandson's) research.

A fundamental issue in vision research is trying to determine how we segregate figure from ground, locating and differentiating the objects from their backgrounds in our environment. Much has been learned since the early work of Gestalt theorists to the present day, yet how this is achieved remains a puzzle. Our understanding of the visual system has grown tremendously since those early days and the tools available to probe this mystery have grown in numbers and technical sophistication, yet we are still left wanting for a complete and clear understanding. In her last published work, Weisstein proposed a model of figure-ground perception based on antagonistic magnocellular (M) and parvocellular (P) pathway interactions in the visual system (Weisstein et al., 1992). The model emerged from research in her lab in the 1980's and was based on what was known at that time about the spatio-temporal frequency and chromatic response properties of the M and P retino-geniculo-cortical pathways. Neural activity initiated in the retina and passed along the M and P pathways is utilized by subsequent brain areas (e.g., in the dorsal and ventral streams) in many different ways. Weisstein's model foreshadowed a number of widely different discoveries related to the dorsal and ventral cortical streams, visual brain physiology and circuitry, and the dynamic, interactive nature of bottom-up and top-down processing. Here we revisit Weisstein's model in light of these different discoveries while also extending it as a general framework for considering eye movement behavior when looking for a figure.

Figure-Ground Perception Based on Antagonistic Dorsal-M & Ventral-P Stream Interactions in The Visual System

At the time Weisstein offered her model there was growing debate about the extent to which the M and P pathways were parallel as well as how and where interactions between them might take place. The M pathway provides the dominant initial feedforward input to the dorsal stream and the P pathway the dominant initial feedforward input to the ventral stream (Haxby et al., 1991; Ungerleider & Haxby, 1994; Ungerleider & Mishkin, 1982). There is now sufficient evidence to argue that antagonistic interactions may be occurring both within the dorsal and ventral streams themselves and between structures in the dorsal and ventral streams (Appelbaum et al., 2006; Doniger et al., 2002; Donner et al., 2008; Wokke et al., 2014). Considering the mixing of M and P connections in V1 and in extrastriate areas found in macaques in both dorsal (Maunsell et al., 1990; Nassi et al., 2006) and ventral streams (Chen et al., 2007; Ferrera et al., 1992, 1994; Merigan & Maunsell, 1993; Nassi & Callaway, 2009; Nealey & Maunsell, 1994; Sincich & Horton, 2005), the neural architecture to afford the dynamic antagonistic interactions of M-dominant dorsal and P-dominant ventral activity Weisstein envisioned would appear to be there. Considering the mixed M and P geniculo-cortical inputs to the dorsal and ventral streams one could describe the inputs to the dorsal stream as dorsal-M and dorsal-P and the inputs to the ventral stream as ventral-P and ventral-M with the relatively dominant inputs being dorsal-M and ventral-P. To best represent the intention of Weisstein's original M-dorsal/P-ventral distinction, we will use the terms dorsal-M and ventral-P when discussing her model and their antagonistic interactions related to figure-ground perception. How then does this relate to our perception of figure and ground?

The basic idea underlying Weisstein's model is that activity in the ventral-P stream encodes figure/foreground information and activity in the dorsal-M stream encodes background information. The ventral-P stream's figure-biased signal is therefore associated with a region being perceived as figure while the dorsal-M stream's ground-biased signal is associated with a region being perceived as ground. In general, where a boundary separates two regions, which region is perceived as figure and which as ground is determined by the outcome of antagonism between dorsal-M and ventral-P signals both within each region and across the boundary between the regions (see Figure 1.1 for an example and Weisstein et al., 1992 for more details). The region with the relatively stronger ventral-P "figure signal" is ultimately perceived as figure while the region with the relatively stronger dorsal-M "ground signal" is perceived as ground (e.g., as represented by the relative circle sizes in Figure 1.1). We next summarize some of the original evidence supporting this idea followed by recent findings consistent with it.

Bottom-Up Influences

One tactic for biasing processing towards dorsal-M vs. ventral-P streams is based on their different spatio-temporal frequency and chromatic sensitivities. The dorsal-M stream is relatively more sensitive to lower spatial and higher temporal frequencies while the ventral-P stream is relatively more sensitive to higher

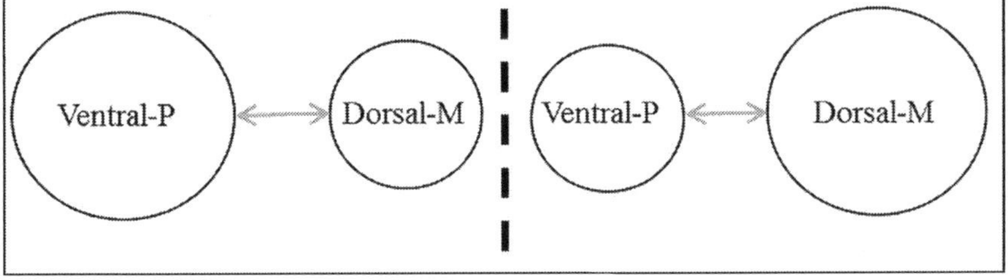

FIGURE 1.1 Figure-ground perception and the spatio-temporal frequency and chromatic response.

Figure-ground perception based on antagonistic dorsal-M & ventral-P stream interactions in the visual system. The perception of a region as figure involves top-down (dashed arrows) and bottom-up influences (circles). Antagonism across the boundary (black double arrow) separating two regions (vertical dashed line) occurs with ventral-P dominant figure and dorsal-M dominant ground signals generated from antagonistic dorsal-M & ventral-P stream interactions (gray arrows) within those regions.

spatial and lower temporal frequencies (Derrington & Lennie, 1984; Kulikowski & Tolhurst, 1973; Legge, 1978; Tolhurst, 1975), and chromatic information (Livingstone & Hubel, 1987, 1988). There is overwhelming evidence indicating spatial frequency differences affect the perception of figure-ground (Klymenko & Weisstein, 1986; Klymenko et al., 1989) and depth (Brown & Weisstein, 1988b). In separate experiments, regions of ambiguous pictures (e.g., the Rubin faces/vase picture (Figure 1.2b), the Maltese Cross (Figure 1.2d), a bipartite field) were filled with sinewave gratings of different spatial frequencies ranging from 0.5 to 8 cpd at one-octave intervals yielding spatial frequency differences ranging from one to four octaves between regions of the ambiguous pictures. The influence of spatial frequency differences on figure-ground perception was measured by the percentage of time regions were perceived as figure. In general, higher spatial frequency filled regions were perceived predominantly as figure relative to lower spatial frequency filled regions, which were perceived as background. This bias to perceive higher spatial frequency filled regions as figure increased as the octave separation between regions increased. These findings also indicated that relative rather than absolute spatial frequency values determine whether a region will be perceived as figure or ground with the perception of figure always biased towards the higher spatial frequency region. The bias to perceive the higher spatial frequency filled regions as figure is accompanied by a strong impression of their being closer in depth (e.g., compare your experience when looking at Figure 1.2c vs 1.2d). This depth impression is so striking that observers consistently place lower spatial frequency filled regions closer in stereo depth to make them appear in the same depth plane as higher spatial frequency filled regions (Brown & Weisstein, 1988b).

There is also substantial evidence indicating temporal frequency differences affect the perception of figure-ground and depth (Klymenko et al., 1989; Wong & Weisstein, 1984, 1985, 1987). These studies used ambiguous figure-ground configurations filled either with random dots (Wong & Weisstein, 1984, 1985, 1987) or sinusoidal gratings (Klymenko & Weisstein, 1989a, 1989b; Klymenko

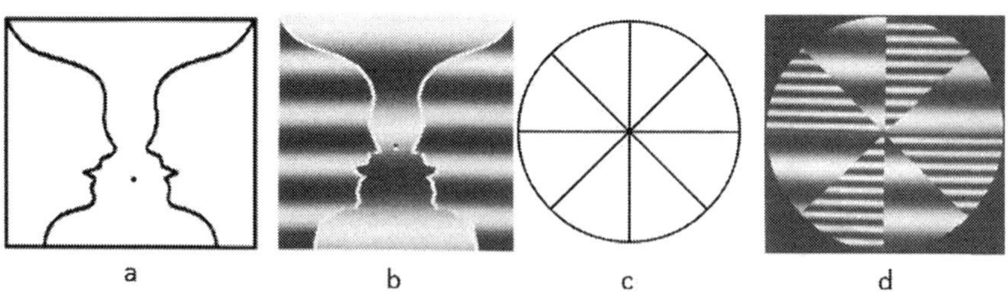

FIGURE 1.2 Figure-ground contexts.

(a, b) Rubin's faces-vase; (c, d) Maltese Cross; (b) and (c) are filled with sinewave gratings as used in experiments examining the effect of spatial frequency on figure-ground perception.

et al., 1989). In general, and particularly for pictures filled with random dots or sinusoidal gratings of the same spatial frequency, flickering (i.e., higher temporal frequency) regions of an ambiguous picture are perceived predominantly as backgrounds, and adjacent non-flickering regions are perceived predominantly as figures closer in depth (also see Palmer & Brooks, 2008, Exp 6).

Single-cell (de Monasterio, 1978; Wiesel & Hubel, 1966), psychophysical (Breitmeyer & Breier, 1994; Breitmeyer & Williams, 1990), and neuroimaging (Bedwell et al., 2006) evidence indicates M-pathway activity is suppressed under diffuse red light conditions. Consistent with the studies reviewed above, under achromatic conditions observers are more likely to perceive the relatively higher spatial frequency-filled half of a bipartite field (1.4 vs. 1.0 cpd) as figure (Weisstein & Brannan, 1991). However, when red and green light matched in luminance is superimposed on opposite sides, the red side is predominantly seen as figure regardless of spatial frequency. Weisstein and Brannan's interpretation was red light decreased dorsal-M activity creating a relatively stronger ventral-P figure signal compared to the green side (also see Palmer & Brooks, 2008, Exp. 3). To summarize, regions of ambiguous figure-ground displays filled with long wavelength red light, a higher spatial frequency and/or a lower temporal frequency are predominantly perceived as figure closer in depth compared to regions filled with another color of equal luminance, a relatively lower spatial frequency and/or higher temporal frequency which are perceived as background, farther away in depth.

Top-Down Influences

The dynamic, interactive nature of ventral-P/dorsal-M stream activity and figure vs. ground perception is also illustrated by how the perception of a region as figure vs. ground influences temporal and spatial processing. The underlying idea is that perceived figure regions bring to bear the relatively slower responding, higher spatial frequency sensitive ventral-P stream while perceived ground regions bring to bear the relatively faster responding, lower spatial frequency sensitive, dorsal-M stream. Evidence of the perception of a region as figure vs. ground influencing temporal processing comes from a masking experiment where a briefly presented (50 ms) mask (i.e., the faces/vase picture in Figure 1.2a) appeared either before, at the same time, or after diagonal target lines appeared at fixation for 50 ms (for details see Weisstein & Wong, 1987). On each trial, observers reported the target line's orientation and then whether they saw the picture as the vase or faces. We focus here only on when the mask appeared after the target (i.e., metacontrast). The underlying assumptions are that both target and mask elicit a fast (M dominant) and slower (P dominant) response, orientation discrimination performance is associated with the slower response to the target, and maximum masking occurs when the fast response to the mask interferes the most with the slower response to the target (see Breitmeyer, 1984; Weisstein, 1972). Longer SOAs in metacontrast imply faster mask processing because mask onset can be delayed and still have its effect on the slower response to the target (e.g., Williams & Weisstein, 1981). Whether the mask was seen as figure or ground, performance was degraded when it appeared after the target, but the stimulus onset asynchrony (SOA) for maximum masking was twice as long for ground compared to figure. When the mask was perceived as the "background boundary" of the region where the target was presented (i.e., when the target was perceived against ground), maximum masking occurred at about 600 ms SOA. When the mask was perceived as the "figure-boundary" of the region where the target was presented (i.e., when the target was perceived against figure), optimal masking was between 200 and 300 ms SOA. Thus, mask processing was faster when it was perceived as ground compared to when perceived as figure.

Examples of perceived figure vs. ground influencing spatial processing comes from findings by Wong and Weisstein (1982, 1983) that sharp-edged target lines (with high spatial frequencies present) are detected better and their orientation discriminated better in perceived figure compared to ground, while blurred target lines (with high spatial frequencies removed) are detected better in perceived ground. Better orientation discrimination of sharp-edged target lines in perceived figure vs. ground has even been found when the perceived figure and ground regions are only in the mind of the observer (i.e., visual phantoms). Visual phantoms involve the perception of a figure/ground distinction in a region of the visual field where none physically exists (see Brown & Weisstein, 1988a, 1991; see also Chapter 4 (Gyoba, Sakurai, & Kitaoka) and 2 (Cavanagh)). Brown and Weisstein (1988a) found better orientation discrimination performance for diagonal line segments flashed in perceived phantom (i.e., figure) vs. non-phantom (i.e., ground) regions of a homogeneous blank region. Importantly, because of the fluctuating nature of visual phantoms, appearing and disappearing over time, this performance difference was only found for trials when observers reported seeing the illusion.

Grounding the Figure: Current Evidence of the Antagonistic Relationship of the Dorsal-M and Ventral-P Streams in Figure-Ground Perception

Before discussing the model further and presenting new evidence in support of it, consider the proposal that the figure "emerges" from the background through a dynamic interaction of activity along and between the P-dominant ventral and M-dominant dorsal streams. One example of evidence for such neural dynamics comes from a study by Applebaum et al. (2006) using an electrophysiological paradigm that allowed them to monitor cortical responses to figure and background separately. They used a frequency tagging procedure where a circular figure was temporally modulated at one frequency relative to a background modulating at a different frequency. They found "the figure region, but not the background, was routed preferentially to lateral cortex. A separate network extending from first tier through more dorsal areas responded preferentially to the background region" (p. 11695).

Another example of dorsal/ventral interactions related to figure-ground segregation comes from two studies by Wokke and colleagues (Wokke et al., 2012, 2014). They used concurrent EEG recordings and TMS to disrupt processing in early visual cortex (i.e., V1/V2) (Wokke et al., 2012, 2014), the lateral occipital cortex (LOC) of the ventral stream, and human MT (HMT) along the dorsal stream (Wokke et al., 2014) while observers performed a motion-defined figure discrimination task. Participants viewed brief exposures of random-dot motion. The dots either all moved in the same direction (homogeneous, i.e., ground), had a square frame of dots (i.e., a figure) moving in a direction different from the background, or had both a frame and its inner square portion moving in different directions from each other and the background (i.e., two "figures/objects" stacked in depth moving against a background). Participants made a three-alternative forced-choice decision which stimulus was presented. Discrimination performance was compared with and without TMS. In both studies, a significant negative deflection in a range between 137–211 ms was found for frame and stack (figure) relative to homogeneous (ground) conditions. When TMS was applied to early visual cortex during an early time window (96–113 ms) this neural correlate of figure-ground segregation was eliminated and performance was significantly impaired. In their 2014 study, TMS was applied over right HMT or LOC 500 ms before stimuli were presented (Exp 1). With TMS to HMT, only stacked performance decreased while the neural correlate of figure-ground segregation in early visual cortex remained. With TMS to LOC, frame and stacked performance increased while the neural correlate of figure-ground segregation also increased. The authors admit their design and methodology did not allow them to determine the precise locus nor direction of dorsal/ventral interactions, but we can interpret their results as indicating a "push-pull interaction in which dorsal and ventral extrastriate areas are being recruited or inhibited depending on stimulus category and task demands" (p. 365) and that these interactions can influence the neural correlate of figure-ground segregation in V1/V2 and the ability to segregate moving stimuli into figure and ground.

A final example of antagonistic dorsal-M/ventral-P neural dynamics directly related to the conscious perception of a region as figure or ground is an fMRI study of motion-induced blindness (MIB) by Donner et al. (2008). They presented a target figure surrounded by a 2° blank region against a background of moving dots that appeared as a rotating sphere which they called a mask. Under steady fixation MIB occurred with the figure periodically disappearing and reappearing over time while observers indicated when it did so. Similar responses were made in a control condition when the target was periodically physically removed and then presented again. By retinotopically locating the target figure relative to the moving background mask in both ventral (V4) and dorsal (MT+, V3AB, pIPS) cortical regions they could monitor responses specific to the target and background mask in those regions when the target was visible vs. invisible and compare them with when the target was physically removed. They found figure region disappearance responses (i.e., when the figure was invisible) in V4 were significantly below baseline during MIB but not when the figure was physically removed.[1] Background mask-specific responses were not found in V4 but were found in areas MT+, V3AB, and pIPS. Opposite of the target-specific responses in V4, mask-specific responses in these dorsal areas increased above baseline during target disappearance and decreased during target reappearance. By contrast, when the target was physically removed, mask-specific responses in these areas tended to decrease.

> The striking dissociation between mask-specific responses during spontaneous disappearance and physical removal of the target is consistent with the hypothesis that cortical representations of the mask in dorsal visual areas (i.e., V3AB and pIPS) play a crucial role in the spontaneous suppression of target representations in ventral visual areas (i.e., V4) during MIB.
>
> *(p. 10305)*

If we consider their results from the perspective of Weisstein's model, the ventral-P dominant figure signal from the target is in an antagonistic relationship with the dorsal-M dominant ground signal from the background regions surrounding the target. During MIB, the target's figure signal is overwhelmed by the surrounding ground signal leading to its disappearance. Dorsal area responses were different when the target was physically removed because there was no longer a figure signal from the target to antagonize against the ground signal.

When Figure-Ground Segregation Fails: New Evidence of the Antagonistic Relationship of the Dorsal-M and Ventral-P Streams in Figure-Ground Perception

The research summarized above provides strong evidence supporting Weisstein's model. Recently we have used an artificial scotoma paradigm as a tool to further explore and test her model. Similar to MIB, the artificial scotoma phenomenon involves a figure region presented in peripheral vision perceptually disappearing with steady fixation even though it remains physically present: a.k.a. "Troxler fading" (Troxler, 1804). From the perspective of Weisstein's model, there is a constant dynamic between ventral-P and dorsal-M activity both within figure and ground regions (Figure 1.3a, light gray arrows) and between them (Figure 1.3a, black arrow). In addition to such Gestalt factors as size and surroundedness, the target in our artificial scotoma experiments is visible as a figure because there is a ventral-P biased figure signal antagonizing against the background's dorsal-M biased ground signal across the border between them. As adaptation of the ventral-P biased figure signal progresses within the figure with steady fixation,

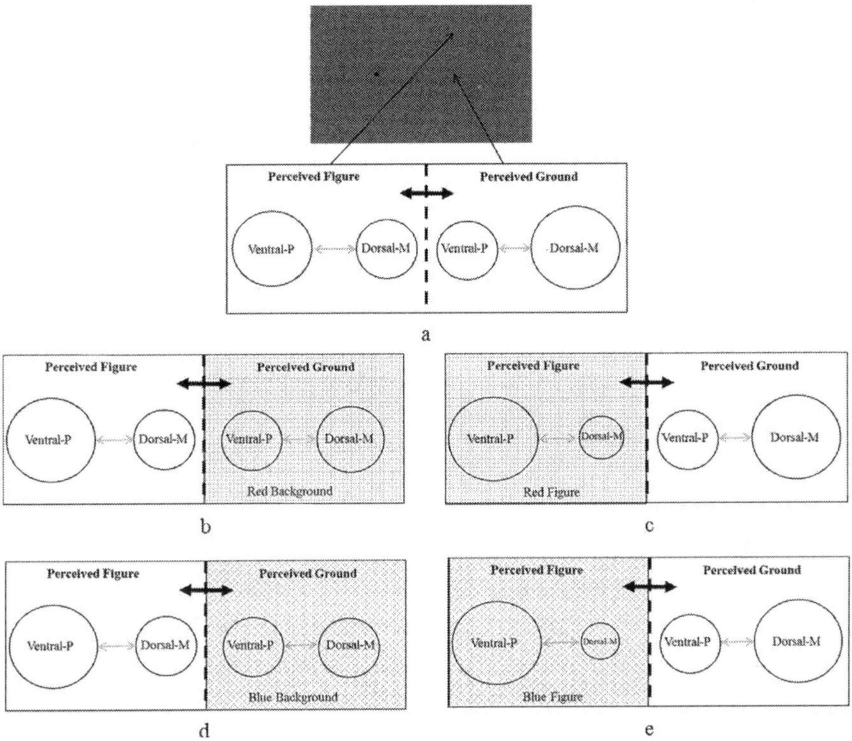

FIGURE 1.3 Balance of antagonistic activity within (gray arrows) and between (black arrows) figure and ground regions.

(a) Example of a same color condition where smaller ventral-P circle in figure represents weaker figure signal (fixation = black dot; blindspot = gray dot). (b–e) Smaller dorsal-M circles in different color conditions represent reduced dorsal-M activity due to red (grid texture) and blue (dot texture) light in figure (c,e) and ground (b,d).

the ability of it to antagonize against (e.g., suppress) the background's dorsal-M biased ground signal declines. Eventually the background's dorsal-M ground signal becomes dominant, suppressing the ventral-P figure signal and allowing the target region to be filled in with the background's dorsal-M biased ground signal (i.e., the target disappears). Being a constantly dynamic process, this antagonism changes over time resulting in the target figure reappearing, similar to how stabilized images fade, reappear, and fade again.

Our strategy was to manipulate dorsal-M activity in both figure and ground regions using red (Exps 1 & 2) and blue (Exp 2) light. As previously noted, single-cell (de Monasterio, 1978; Wiesel & Hubel, 1966), psychophysical (Breitmeyer & Breier, 1994; Breitmeyer & Williams, 1990), and neuroimaging (Bedwell et al., 2006) evidence indicates dorsal-M activity is suppressed under red light. With no

(Dacey, 2000; Sun et al., 2006a, 2006b) or minimal (Chatterjee & Callaway, 2002) blue cone input to M ganglion cells, M activity was expected to be dramatically reduced under blue light. This might explain why fade times were longest for blue compared to other colors tested in the lone artificial scotoma study examining the influence of color we could find (Sakaguchi, 2001). Depending on whether red or blue light appears in the figure or ground region, it would be expected to alter the dorsal-M/ventral-P balance of activity within the region producing a shift in the balance of activity across the border between figure and ground. Based on this strategy we made the following predictions illustrated in Figure 1.3b–e, where smaller dorsal-M labelled circles represent reduced dorsal-M activity shifting the balance towards relatively greater ventral-P activity. First, red (b) or blue (d) light filling the ground should reduce dorsal-M activity shifting the balance of dorsal-M/ventral-P activity there. This shift in the balance away from dorsal-M activity should weaken the ground signal making it less able to overwhelm the figure signal resulting in longer fade times. Conversely, red (c) or blue (e) light in the figure should reduce dorsal-M activity there, shifting the balance even further towards relatively greater ventral-P activity resulting in a stronger figure signal. A stronger figure signal should make the figure more resistant to being overwhelmed by the ground signal resulting in longer fade times.

In our experiments the figure region was a 2° × 2° homogeneous square presented against a random-dot background subtending 26.7° (w) × 17.2° (h). The figure was located 10° from fixation and presented randomly in either the top right or bottom right of the display (see Figure 1.3a for an example). Viewing was monocular with the right eye. A blindspot stimulus was used to monitor eye movements and any trials observers indicated it became visible were removed. In Experiment 1 the background either flickered at 14 frames per second or was static and in Experiment 2 it was only static. Background flicker had no effect so only static results are presented. Observers pressed one key to indicate the figure had completely faded and another key when the figure reappeared. Trial duration was 30 sec, terminating if a reappearance response was made or self-terminating if no response was made. The random-dot backgrounds were composed of equal amounts of two different luminance levels of the same color, with a mean luminance of either 25 cd/m^2 (Exp 1, N=15) or 3 cd/m^2 (Exp 2, N=7)[2] with the figure always physically equiluminant with the background. In Experiment 1 every possible figure/background combination of gray (Gy), green (Gn), and red (Rd) were tested, resulting in three same (Gy/Gy, Gn/Gn, Rd/Rd) and six different color combinations (Gy/Rd, Gy/Gn, Gn/Gy, Gn/Rd, Rd/Gy, Rd/Gn). The use of blue in Experiment 2 resulted in a fourth same color (Bl/Bl) and six additional different color combinations (Bl/Gy, Gy/Bl, Bl/Gn, Gn/Bl, Bl/Rd, Rd/Bl).

The short fade times in the same color conditions (Figures 1.4a, 1.5a, b) indicate a weak figure signal (Figure 1.3a) being overwhelmed most easily by the ground signal. This occurs because only a lack of texture defines the figure making ventral-P activity most similar in figure and ground. Fade times are longer in the different color conditions because a stronger ventral-P figure signal is produced by the color difference making it more resistant to the ground signal. These conditions allowed us to reveal how figure-ground segregation is influenced by manipulations of dorsal-M activity by the changes in fade times depending on the color of the figure and ground regions.

To evaluate the influence of color differences devoid of the texture difference between figure and ground we used the fade times from same color conditions as baselines and subtracted them from different color conditions fade times. Consider the following example using Gy/Gy. The logic is Gy/Gy fade times are due to the influence of Gy in the figure and Gy in the ground and the texture difference between them. By subtracting Gy/Gy fade times from different color condition fade times containing Gy we obtain a measure of the influence of the other color. For example, by subtracting Gy/Gy from Gy/Gn fade times we are left with the effect of a green ground when the figure was gray. Similarly, by subtracting Gy/Gy from Gy/Rd fade times we are left with the effect of a red ground when the figure was gray. We can now directly compare green vs. red as ground when the figure was gray (e.g., see far left panel of Figure 1.4). This

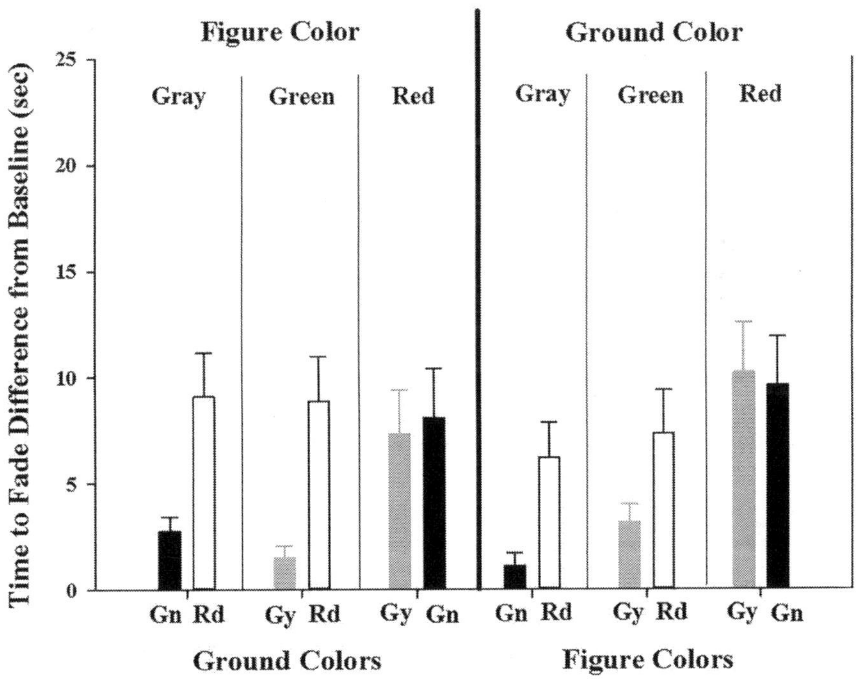

FIGURE 1.4 The time it took for a 2° square figure presented 10° from fixation to disappear after subtracting the same color condition fade time as baseline (see text for details).

was done for all different color combinations tested and are shown in Figure 1.4 (Exp 1) and Figure 1.5a, b (Exp 2).

Repeated-measures ANOVAs on fade times across color combinations revealed significant main effects of color for both Experiment 1 [$F(8,112) = 11.99$, $p < .0001$] and Experiment 2 [$F(15,90) = 17.11$, $p < .0001$]. Planned pairwise comparisons found no difference in fade times across the same color combinations and these conditions all had shorter fade times than all the different color combinations. As is clear in Figure 1.4, differences in fade time from baseline were similar when red was either figure or ground (third and sixth panels from the left). Separate 2 (figure vs ground) × 2 (color) repeated measures ANOVAs were used to compare fade times for green vs. red as either figure or ground with Gy/Gy subtracted as baseline (Figure 1.4, first and fourth panels from the left), and for gray vs. red as either figure or ground with Gn/Gn subtracted as baseline (Figure 1.4, second and fifth panels from the left). The first ANOVA with Gy/Gy subtracted as baseline revealed significant main effects of figure/ground [$F(1,14) = 5.67$, $p < .032$] and color [$F(1,14) = 13.44$, $p < .003$] and no interaction. Fade times were longer for (1) ground (5.9 sec) than for figure (3.6 sec) (i.e., the average bar height in Figure 1.4 is greater in the fourth panel from the left compared to the first panel) and (2) for red (7.6 sec) than for green (1.9 sec) (i.e., the white vs. black bars in those panels). The second ANOVA with Gn/Gn subtracted as baseline found only a main effect of color [$F(1,14) = 14.93$, $p < .002$] with red fade times longer (8.1 sec) than gray (2.3 sec) irrespective of whether the region was figure or ground.

The pattern of results for Experiment 2 for color combinations not containing blue were generally very similar to Experiment 1 as evident by visual inspection of Figures 1.4 and 1.5. While not all planned pairwise comparisons significant in Experiment 1 were significant in Experiment 2, this is likely due to fewer observers. Similar to Experiment 1, there were no differences in fade times across the Gy/Gy, Gn/Gn, and Rd/Rd conditions. Bl/Bl fade times were shorter than Gn/Gn ($p < .03$) and all different color combinations. Gy/Gn and Gn/Gy fade times were shorter than all other combinations with red and all combinations with blue. Finally, there were many color combinations with blue (Gy/Bl, Gn/Bl, Bl/Gn, Rd/Bl, Bl/Rd) with significantly longer fade times than color combinations with red (Gy/Rd, Rd/Gy) (all p values $\leq .04$). Overall these findings support the predictions based on Weisstein's model due to reduced dorsal-M activity under red and blue light conditions.

As is clear in Figure 1.5a & b (right most panels), differences in fade time from baseline were the largest and similar whether blue was figure or ground. Separate 2 (figure vs ground) × 3 (color) repeated measures ANOVAs were used to compare fade times for green vs. red vs. blue as either figure or ground with Gy/Gy subtracted as baseline (leftmost panels in Figure 1.5a & b), for gray vs. red vs. blue as either figure or ground with Gn/Gn subtracted as baseline (second panels from the left in Figure 1.5a & b), and for gray vs. green vs. blue as either figure or ground

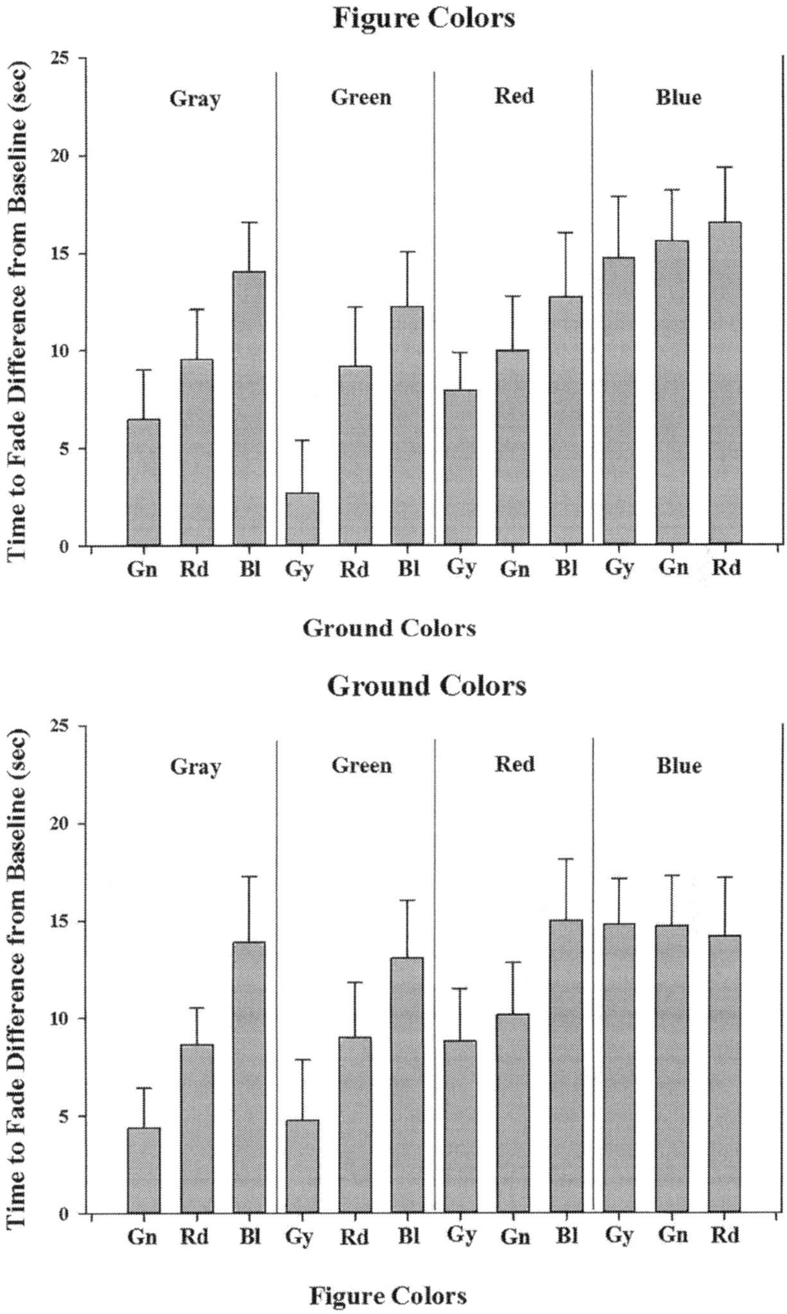

FIGURE 1.5 The time it took for a 2° square figure presented 10° from fixation to disappear after subtracting the same color condition fade time as baseline.

with Rd/Rd subtracted as baseline (third panels from the left Figure 1.5a & b). Only significant main effects of color were found for each ANOVA (Gy/Gy [$F(2,12) = 9.15$, $p < .004$], Gn/Gn [$F(2,12) = 18.18$, $p < .0001$], and Rd/Rd [$F(2,12) = 6.17$, $p < .014$]). Probing these main effects, with Gy/Gy subtracted as baseline, blue fade times (13.96 sec) were longer than green (5.42 sec) and red (9.09 sec); with Gn/Gn subtracted as baseline, blue fade times (12.65 sec) were longer than gray (3.71 sec) and red (9.07 sec); with Rd/Rd subtracted as baseline, blue fade times (13.84 sec) were longer than gray (8.37 sec) but not green (10.07 sec). These and the time-to-fade results support our predictions related to blue light.

As a way of exploring and testing Weisstein's model of figure-ground perception based on antagonistic dorsal-M and ventral-P stream interactions we used a phenomenon where figure-ground segregation fails. Using equiluminant stimuli and red and blue light expected to reduce dorsal-M activity, the specific predictions made based on her model were largely born out. Reducing dorsal-M activity in the ground weakened the ground signal making it less able to overwhelm the figure signal resulting in longer fade times. Conversely, reducing dorsal-M activity in the figure strengthened the figure signal making it more resistant to being overwhelmed by the ground signal resulting in longer fade times. Clearly these results have similarities with and implications for, and need to be considered from the perspective of, other models such as Peterson's Parallel Interactive Model (Peterson & Cacciamani, 2013; Peterson et al., 2000), Vecera's biased-competition model (Vecera, 2000), Grossberg's model (Grossberg, 1994, 2016) and others, particularly those emphasizing boundary ownership (Kogo & van Ee, 2015; von der Heydt, 2015), but such an in depth discussion is beyond the scope of this chapter. We next turn to how Weisstein's thinking has influenced our research on eye movement behavior.

When Do You Look Where You Look? Antagonism in Searching For a Figure

Weisstein's model of M-P antagonism has also inspired and informed our work on determining the nature of eye movement behavior when one searches for a figure in the environment. Our recent work has examined how fixation behavior, in particular presaccadic fixation durations, may be influenced by such antagonism. Spatially-distributed information in retinal images benefit from detailed (i.e., P-pathway dominant) processing only within the fovea, a region which spans about 2 deg. of visual angle. As most of the signals impinging on the retina are not accessible to detailed processing, fast eye movements (i.e., saccades) have to be executed to bring points of interest into foveal processing range. Visual awareness is mostly suppressed during saccades (Bremmer et al., 2009; Matin, 1974; Wurtz, 2008) so useful information processing (e.g., figure identification) starts only when the eyes are relatively stable (i.e., fixed to a point of interest). A saccade is initiated when a potential figure, located and segmented from the ground in peripheral vision, is selected for detailed foveal processing.

Saccades are controlled by a network of cortical and subcortical areas, through the superior colliculus in the mid brain (see Goldberg & Walker, 2013; White & Munoz, 2011 for reviews). Physiological studies of nonhuman primates suggest that with respect to retinal input, the SC is innervated by a variety of retinal ganglion cells that make up the koniocellular (K) pathway (May, 2014). The K pathway is much less understood than the M and P pathways, but has properties which are similar to cells in either the M or the P pathway (Hendry & Reid, 2000; Kaplan, 2014). The K pathway contributes both to foveal vision (which presumably involves figure processing) and to peripheral vision (see Masri et al., 2016). Nonhuman primate physiology of the SC also indicates that cortical afferent inputs are evident from the visual cortex, the M-dominated dorsal lobes (e.g., dorsolateral prefrontal area, parietal eye fields, supplementary and frontal eye fields), and to a smaller extent, the P-dominated temporal lobe (Baldwin & Kaas, 2012; Cerkevich et al., 2014; Fries, 1984; Munoz & Schall, 2004). Thus to date, there is evidence to suggest that the SC is strongly innervated by the dorsal-M pathway, and to some extent by the K, and the ventral-P pathways. The combination of inputs to the SC may account for recent exciting findings about the human SC. Reports indicate that the human SC exhibits not only typical color-blind, fast attention-switching M-dominant responses, but also color contrast (Chang et al., 2016), and complex figure processing responses (Georgy et al., 2016; Nguyen et al., 2014) typically attributed to P-dominance. While the traditional approach which tightly links specific pathways to specific functions may be overly simplistic (see Freud et al., 2016; Kaplan, 2014)—a theme that runs through Weisstein's work—relativity in antagonism, appears to be preserved in the SC. In the remaining sections we show that antagonism is apparent in the initiation of saccades during visual search.

Eye Fixation and Saccade Zones

Cells in the rostral and caudal areas of the SC behave differently (see Gandhi & Katnani, 2011; Munoz & Fecteau, 2002 for reviews). In rostral areas, cells are vigorously active for the execution of small-amplitude saccades. Towards the caudal areas in contrast, cells are increasingly active for the execution of larger amplitude saccades. This organization suggests the existence of two functional zones in the SC (Munoz & Fecteau, 2002)—a fixation zone (optimized for small amplitude saccades within a narrow area of a figure), and a saccade zone (optimized for re-orienting attention with larger saccades in search of parafoveal/peripheral information). The two functional zones are of interest for our psychophysics work on determining how we search with eye movements for a figure in the environment. In brief, the fixation zone (which in the present context, we suggest is optimized for attending to figures) is innervated by foveal projections, and the saccade zone (which we suggest is optimized for attending to ground) is innervated by the retinal periphery (see Munoz & Schall, 2004 for a discussions of SC connections). The balance between eye fixations and saccade execution depends on antagonistic interactions between the two zones (e.g., Munoz & Fecteau, 2002). During an eye fixation, fixation zone activity inhibits more distant saccade zone activity. A voluntary saccade is initiated towards a potential figure when a saccadic decision made in the cortex pushes saccade zone activity to overcome fixation zone activity (Munoz & Fecteau, 2002; Munoz & Wurtz, 1993). The antagonistic interaction between fixation and saccadic zones has been used to explain saccade latencies (see Munoz & Fecteau, 2002; Olivier et al., 1999) as latencies are generally longer when the balance of activity is not overwhelmingly on the side of the saccade zone.

Vertical Visual Field Asymmetry in Saccade Timing

From rostral to caudal areas in the SC, medially located cells are optimized for saccades with upward vectors, and laterally located cells are optimized for saccades with downward vectors (e.g., Gandhi & Katnani, 2011). It has also been shown that compared to SC cells responding to peripheral targets in the lower visual field (LoVF), cells responding to peripheral targets in the upper visual field (UpVF) exhibit shorter response latencies (Hafed & Chen, 2016). Beyond the SC, human magnetoencephalography (MEG) work has shown that frontal lobe areas involved in saccade preparation are accessed earlier and with less cortical activation/effort when a saccade is about to be directed into the UpVF (Tzelepi et al., 2010). Shorter response latencies in the UpVF of the SC and the temporal advantage in the cortex for upward-bound saccades may account for human and monkey psychophysical findings of faster saccade reaction times (SRTs) for spot stimuli presented in the UpVF (Abegg et al., 2015; see also Greene et al., 2014 for a brief review). Timing mechanisms in the SC and cortex may also explain repeated findings of shorter presaccadic fixation durations (PSFDs) during visual search, before saccades are directed upwards (Greene & Brown, 2017; Greene et al., 2014). Typical PSFD findings are illustrated in Figure 1.6a (Experiment 1 from Greene et al., 2014). Given physiological findings to date (Hafed & Chen, 2016; Munoz & Fecteau, 2002; Tzelepi et al., 2010), the vertical asymmetry may reasonably be explained by a stronger pull from the saccade zone in the UpVF, making it easier to overcome the fixation zone when an upward-bound saccade is intended. With knowledge only of antagonistic interactions between saccade and fixation zones (e.g., Munoz & Fecteau, 2002) and the MEG findings of Tzelepi et al. (2010), Greene and Brown (2017) tested the stronger UpVF saccade zone explanation using a gaze-contingent task. An eye tracker was programmed to obstruct peripheral information (see Figure 1.6b) as a function of the observer's direction of gaze on a visual display. The task was referred to as

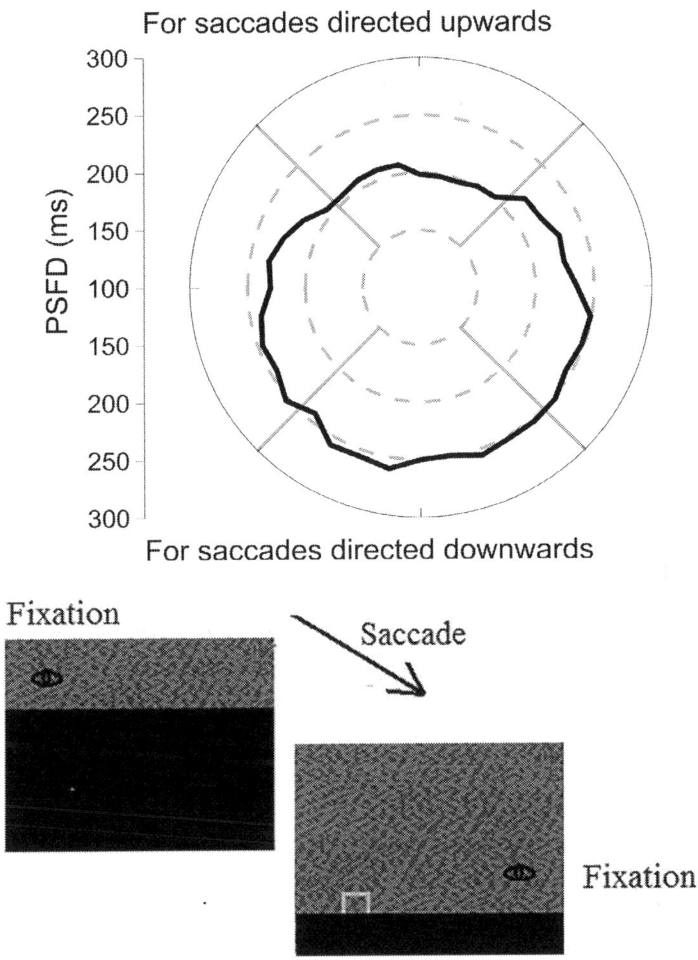

FIGURE 1.6 Polar plots depicting presaccadic fixation durations (PSFDs) in various visual search experiments.

In all cases, PSFDs were shorter before saccades were directed upwards than downwards. In (a) search was performed with no constraints. [Adapted from Greene et al., 2014, with permission from Elsevier.] (b) Illustration of a participant's perspective for a display with a gaze-contingent obstruction (i.e., black area) in the lower visual field. Here, the participant who first looks to the upper left of the display (see the eye), and then to the lower right. [Adapted from Greene & Brown, 2017.] In (c) a gaze-contingent moving obstruction removed visibility of information in the upper, or the lower visual field as search proceeded. The prolongation of PSFDs for upward-bound saccades when there was an obstruction in the UpVF was less than the prolongation of PSFDs for downward-bound saccades when there was an obstruction in the LoVF. This suggests greater resistance to prolongation of PSFDs in the UpVF. [Adapted from Greene & Brown, 2017.] In (d) PSFDs were compared in red and grey search displays. During search of the red relative to the grey display, the prolongation of PSFDs for upward-bound saccades was less than the prolongation of PSFDs for downward-bound saccades. This suggests greater resistance to prolongation of PSFDs in the UpVF. Together, the findings suggest that the UpVF is less susceptible to attenuation of saccade zone mechanisms than the LoVF.

(Continued)

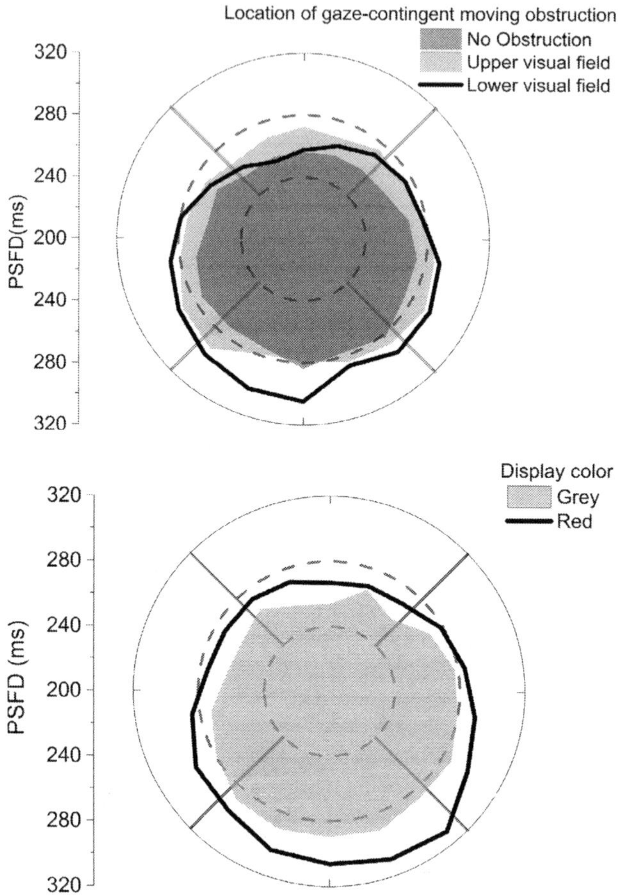

FIGURE 1.6 Continued

a gaze-contingent moving obstruction task (see Rayner, 1998 for similar gaze-contingent descriptions).

Reducing Saccade Zone Antagonism by Obstructing Peripheral Preview

Greene and Brown (2017) reasoned that if the SC fixation and saccade zones interact antagonistically (e.g., Munoz & Fecteau, 2002), and if upward-bound saccades require less cortical effort (Tzelepi et al, 2010), then there are predictable behavioral effects of manipulating the zones. Suppression of the saccade zone should lead to increased PSFD overall. However, given a hypothesized stronger saccade zone in the UpVF, greater resistance to suppression of the SC saccade zone was expected when the UpVF was obstructed. Such a resistance to suppression should make it difficult for PSFD to increase from baseline. In other words, saccade zone suppression was expected to have less of an impact on PSFDs for upward-bound than downward-bound saccades in a visual search task. A gaze-contingent moving obstruction task was devised, based on findings about the behavior of saccade-related neurons in the SC.

Extracellular recordings from saccade-related neurons in the SC of adult rhesus monkeys indicate that neurons are most active when saccades are directed to visible, locally-defined peripheral targets (Edelman & Goldberg, 2001, 2003; Marino et al., 2012). Reasonably, if peripheral information is obstructed during visual search of a random dot display, this should reduce presaccadic activity towards the obstructed region. Greene and Brown (2017) hypothesized that a peripheral gaze-contingent moving obstruction in the UpVF or LoVF should reduce the activity of the relevant saccade zone, thus increasing PSFDs in a direction selective manner. As predicted, in the presence of a location-specific peripheral gaze-contingent moving obstruction, PSFDs were significantly prolonged *only if a saccade was executed in the direction of the obstruction* (as shown in Figure 1.6c). Control experiments with peripheral gaze-contingent moving lines confirmed that the effects found may not be attributed to moving stimulation in the peripheral visual field (Greene & Brown, unpublished data), but to the moment-by-moment covering of peripheral information. The results may be interpreted as evidence for a suppression of saccade zone activity by the obstruction in the direction of upcoming saccades. This suppression led to the maintenance of direction-specific fixation zone activity and hence, longer direction-specific PSFDs. More importantly, as might be expected from a theory positing stronger antagonism/resistance to suppression in the UpVF, the effect size was smaller for saccades moving into the obstruction in the UpVF (Tukey HSD adjusted $p < .01$; $d = .58$) than for saccades moving into the obstruction in the LoVF (Tukey HSD adjusted $p < .01$; $d = .80$). The findings are consistent with Hafed and Chen's (2016) claim of a sharper, stronger, faster UpVF representation in primate SC. In the present context, we have also interpreted this to mean the saccade zone exhibits stronger antagonism in the UpVF. We have tested the saccade zone suppression hypothesis further, using diffuse red light in a visual search task.

Reducing Saccade Zone Antagonism with Diffuse Red Light

Given the previously noted neurophysiological and psychophysical evidence that diffuse red light suppresses M-dominated activity, we reasoned that a red background would suppress the saccade zone. This was predicated on two sets of research findings. First, covert and overt attention share the M-dominated parieto-frontal cortical network (see Goldberg & Walker, 2013; Petersen & Posner, 2012; Schiller & Tehovnik, 2009), with this network providing the bulk of the input to the SC (e.g., Munoz & Schall, 2004). Second, our recent work (Greene & Brown,

2017) suggested that saccade-zone antagonism may be attenuated by experimental manipulation. Using a visual search task similar to Greene and Brown (2017), but this time with no obstruction, participants searched for the target in physically equiluminant gray vs. red random-dot displays in counter-balanced blocks of trials. Given the suggested stronger saccade zone in the UpVF (e.g. Hafed & Chen, 2016; Greene & Brown, 2017; Tzelepi et al., 2010), we expected more resistance to suppression in the UpVF. In other words, the increase in PSFDs due to the attenuation of M-dominated activity was expected to be less in the UpVF than the LoVF. PSFD results from this experiment are shown in Figure 1.6d. The important point made by this experiment is not that the vertical asymmetry reported earlier (Greene et al., 2014; Greene & Brown, 2017) was replicated, or that PSFDs were longer for the red than gray display - probably due to saccade zone attenuation by diffuse red light. The primary concern was whether or not the UpVF would be more resistant to suppression. The expectation was that across the two types of displays (red vs gray), PSFDs for saccades into the UpVF would be more alike than PSFDs for saccades into the LoVF. *A priori* analyses supported our expectations. For upward-bound saccades, with (conservative) Bonferroni adjustments, there was no significant difference in PSFDs between the red and gray search displays (Bonferroni adjusted $p = .12$; $d = .79$). In effect, the UpVF was rather resistant to saccade zone suppression, and PSFDs did not significantly increase. In contrast, for downward-bound saccades, PSFDs were significantly longer for the red display (Bonferroni adjusted $p = .01$; $d = 1.15$), suggestive of a strong attenuation of saccade zone antagonism in the LoVF. The findings support a theory of stronger resistance to saccade antagonism in the UpVF.

When You Look Where You Look — a speculative account of UpVF and LoVF differences

We speculate that an adaptive "check and detect" behavior pattern may underlie these PSFD findings (and those of Greene et al., 2014 and Greene et al., 2017) and the suggested antagonism in the saccade circuitry. Given the position of the eyes atop the human body, distant objects tend to be in the UpVF, and objects near the body, in the LoVF. As one takes a walk in the park, for example, gaze is typically directed some distance ahead to search for, and detect information. Only occasionally, because one is covertly aware of near-body/peripersonal space, does one look down, typically to check for salient information in one's path (e.g., Is that a branch or snake near one's foot?). In effect, peripersonal space requires only infrequent eye movement checks (see also Greene, et al., 2014 for probability of visual search saccades). Far/extrapersonal space in contrast, requires many eye movements to detect the multitude of information beyond one's personal space. Attending to peripersonal space requires downward-directed saccades, and attending to extrapersonal space requires upward-directed saccades (see Previc, 1990). The need for upward-directed saccades during visual search of extrapersonal space may underlie the fast representation of saccades intended for the UpVF (as reported by Hafed & Chen, 2016). Higher spatial frequency tuning in the UpVF (as reported by Hafed & Chen, 2016) may facilitate this important adaptation to direct saccades upwards in search of task-relevant forms (i.e., potential figures) in extrapersonal space. For example, as one's walk in the park continues, it may be important for saccade selection to distinguish between the features of a warning sign (i.e., diamond shaped object on a post) and a small tree in extrapersonal space.

Naomi Weisstein's findings on relative antagonism in visual information processing contributed to how we have hypothesized about, and interpreted eye movement behavior in our collaborative efforts. Perhaps, a more general antagonism is between where we are fixating (incorporating *what is it* mechanisms) and where we are about to look (incorporating *where is it/how to utilize it* mechanisms). In this chapter, Naomi Weisstein's antagonistic theme, and the traditional dichotomies of visual information processing (e.g., Milner & Goodale, 2007; Ungerleider & Mishkin, 1982) are generalized to a context that involves visual brains "reaching out" to engage figure/ground relationships in the world. Inspired by Professor Naomi Weisstein, we are continuing to study the mind!

Notes

1 The authors "focused on the disappearance responses to distinguish response decreases (that we interpret as being involved in target perception) and response increases (that are confounded with attention capture)" (Donner et al, 2008, p. 10303).
2 The reduced luminance range for blue required scaling all other stimuli to its maximum luminance.

References

Abegg, M., Pianezzi, D., & Barton, J. (2015). A vertical asymmetry in saccades. *Journal Of Eye Movement Research*, (5), 1–10.

Appelbaum, L. G., Wade, A. R., Vildavski, V. Y., Pettet, M. W., & Norcia, A. M. (2006). Cue-invariant networks for figure and background processing in human visual cortex. *The Journal of Neuroscience, 26*(45), 11695–11708.

Baldwin, M. K. L., & Kaas, J. H. (2012). Cortical projections to the superior colliculus in prosimian galagos (Otolemur garnetti). *The Journal of Comparative Neurology, 520*(9), 2002–2020.

Bedwell, J. S., Miller, L. S., Brown, J. M., & Yanasak, N. E. (2006). Schizophrenia and red light: fMRI evidence for a novel biobehavioral marker. *International Journal of Neuroscience, 116*(8), 881–894.

Breitmeyer, B. G. (1984). *Visual Masking: An Integrative Approach*. Oxford: Clarendon.

Breitmeyer, B. G., & Breier, J. I. (1994). Effects of background color on reaction time to stimuli varying in size and contrast: inferences about human M channels. *Vision Research, 34*(8), 1039–1045.

Breitmeyer, B. G., & Williams, M. C. (1990). Effects of isoluminant-background color on metacontrast and stroboscopic motion: Interactions between sustained (P) and transient (M) channels. *Vision Research, 30*(7), 1069–1075.

Bremmer, F., Kubischik, M., Hoffmann, K.-P., & Krekelberg, B. (2009). Neural dynamics of saccadic suppression. *The Journal of Neuroscience : the Official Journal of the Society for Neuroscience, 29*(40), 12374–12383.

Brown, J. M., & Weisstein, N. (1988a). A phantom context effect: Visual phantoms enhance target visibility. *Perception & Psychophysics*, *43*(1), 53–56.

Brown, J. M., & Weisstein, N. (1988b). A spatial frequency effect on perceived depth. *Perception & Psychophysics*, *44*(2), 157–166.

Brown, J. M., & Weisstein, N. (1991). Conflicting figure–ground and depth information reduces moving phantom visibility. *Perception*, *20*(2), 155–165.

Cerkevich, C., Lyon, D., Balaram, P., & Kaas, J. (2014). Distribution of cortical neurons projecting to the superior colliculus in macaque monkeys. *Eye and Brain*, *6*, 121–137.

Chang, D. H. F., Hess, R. F., & Mullen, K. T. (2016). Color responses and their adaptation in human superior colliculus and lateral geniculate nucleus. *NeuroImage*, *138*, 211–220.

Chatterjee, S., & Callaway, E. M. (2002). S cone contributions to the magnocellular visual pathway in macaque monkey. *Neuron*, *35*(6), 1135–1146.

Chen, C.-M., Lakatos, P., Shah, A. S., Mehta, A. D., Givre, S. J., Javitt, D. C., & Schroeder, C. E. (2007). Functional anatomy and interaction of fast and slow visual pathways in macaque monkeys. *Cerebral Cortex*, *17*(7), 1561–1569.

Dacey, D. M. (2000). Parallel pathways for spectral coding in primate retina. *Annual Review of Neuroscience*, *23*(1), 743–775.

de Monasterio, F. M. (1978). Properties of concentrically organized X and Y ganglion cells of macaque retina. *Journal of Neurophysiology*, *41*(6), 1394–1417.

Derrington, A. M., & Lennie, P. (1984). Spatial and temporal contrast sensitivities of neurones in lateral geniculate nucleus of macaque. *The Journal of Physiology*, *357*(1), 219–240.

Doniger, G. M., Foxe, J. J., Murray, M. M., Higgins, B. A., & Javitt, D. C. (2002). Impaired visual object recognition and dorsal/ventral stream interaction in schizophrenia. *Archives of General Psychiatry*, *59*(11), 1011–1020.

Donner, T. H., Sagi, D., Bonneh, Y. S., & Heeger, D. J. (2008). Opposite neural signatures of motion-induced blindness in human dorsal and ventral visual cortex. *The Journal of Neuroscience*, *28*(41), 10298–10310.

Edelman, J. A., & Goldberg, M. E. (2001). Dependence of saccade-related activity in the primate superior colliculus on visual target presence. *Journal of Neurophysiology*, *86*(2), 676–691.

Edelman, J. A., & Goldberg, M. E. (2003). Saccade-related activity in the primate superior colliculus depends on the presence of local landmarks at the saccade endpoint. *Journal of Neurophysiology*, *90*(3), 1728–1736.

Ferrera, V. P., Nealey, T. A., & Maunsell, J. H. R. (1992). Mixed parvocellular and magnocellular geniculate signals in visual area V4. *Nature*, *358*(6389), 756–758.

Ferrera, V. P., Nealey, T. A., & Maunsell, J. H. R. (1994). Responses in macaque visual area V4 following inactivation of the parvocellular and magnocellular LGN pathways. *The Journal of Neuroscience*, *14*(4), 2080–2088.

Freud, E., Plaut, D. C., & Behrmann, M. (2016). 'What' is happening in the dorsal visual pathway? *Trends in Cognitive Sciences*, *20*(10), 773–784.

Fries, W. (1984). Cortical projections to the superior colliculus in the macaque monkey: A retrograde study using horseradish peroxidase. *The Journal of Comparative Neurology*, *230*(1), 55–76.

Gandhi, N. J., & Katnani, H. A. (2011). Motor functions of the superior colliculus. *Annual Review of Neuroscience*, *34*, 205–231.

Georgy, L., Celeghin, A., Marzi, C. A., Tamietto, M., & Ptito, A. (2016). The superior colliculus is sensitive to gestalt-like stimulus configuration in hemispherectomy patients. *Cortex*, *81*, 151–161.

Goldberg, M. E., & Walker, M. F. (2013). The control of gaze. In E. R. Kandel, J. H. Schwartz, T. M. Jessell, S. A. Siegelbaum, & A. J. Hudspeth (Eds), *Principles of Neural Science* (pp. 894–916). New York: McGraw-Hill.

Greene, H. H., & Brown, J. M. (2017). Where did I come from? Where am I going? Functional differences in visual search fixation duration. *Journal of Eye Movement Research, 10*(1), 1–13.

Greene, H. H., Brown, J. M., & Dauphin, B. (2014). When do you look where you look? A visual field asymmetry. *Vision Research, 102*, 33–40.

Grossberg, S. (1994). 3-D vision and figure-ground separation by visual cortex. *Perception & Psychophysics, 55*(1), 48–120.

Grossberg, S. (2016). Cortical dynamics of figure-ground separation in response to 2D pictures and 3D scenes: How V2 combines border ownership, stereoscopic cues, and Gestalt grouping rules. *Frontiers in Psychology, 6*.

Hafed, Z. M., & Chen, C.-Y. (2016). Sharper, stronger, faster upper visual field representation in primate superior colliculus. *Current Biology, 26*(13), 1647–1658.

Haxby, J. V., Grady, C. L., Horwitz, B., Ungerleider, L. G., Mishkin, M., Carson, R. E., Herscovitch, P., Schapiro, M. B., & Rapoport, S. I. (1991). Dissociation of object and spatial visual processing pathways in human extrastriate cortex. *Proceedings of the National Academy of Sciences, 88*(5), 1621–1625.

Hendry, S. H. C., & Reid, R. C. (2000). The koniocellular pathway in primate vision. *Annual Review of Neuroscience, 23*(1), 127–153.

Hoffman, J. (1996). Visual object recognition. In W. Prinz & B. Bridgeman (Eds), *Perceptual Organization* (pp. 297–344). New York: Academic Press.

Kaplan, E. (2014). The M, P, and K pathways of the primate visual system revisited. In W. C. Hall & A. Moschovakis (Eds), *The New Visual Neurosciences* (pp. 215–226). London: MIT Press.

Klymenko, V., & Weisstein, N. (1986). Spatial frequency differences can determine figure-ground organization. *Journal of Experimental Psychology Human Perception & Performance, 12*(3), 324–330.

Klymenko, V., & Weisstein, N. (1989a). Figure and ground in space and time: 1. Temporal response surfaces of perceptual organization. *Perception, 18*(5), 627–637.

Klymenko, V., & Weisstein, N. (1989b). Figure and ground in space and time: 2. Frequency, velocity, and perceptual organization. *Perception, 18*(5), 639–648.

Klymenko, V., Weisstein, N., Topolski, R., & Hsieh, C. H. (1989). Spatial and temporal frequency in figure-ground organization. *Perception & Psychophysics, 45*(5), 395–403.

Kogo, N., & van Ee, R. (2015). Neural mechanisms of figure-ground organization: Border-ownership, competition and perceptual switching. In J. Wagemans (Ed.), *The Oxford handbook of perceptual organization*. Oxford: Oxford University Press.

Kulikowski, J. J., & Tolhurst, D. J. (1973). Psychophysical evidence for sustained and transient detectors in human vision. *Journal of Physiology, 232*(1), 149–162.

Legge, G. E. (1978). Sustained and transient mechanisms in human vision: Temporal and spatial properties. *Vision Research, 18*(1), 69–81.

Livingstone, M. S., & Hubel, D. H. (1987). Psychophysical evidence for separate channels for the perception of form, color, movement, and depth. *Journal of Neuroscience*, 7(11), 3416–3468.

Livingstone, M. S., & Hubel, D. H. (1988). Segregation of form, color, movement, and depth: Anatomy, physiology, and perception. *Science, 240*(4853), 740–749.

Marino, R. A., Levy, R., Boehnke, S., White, B. J., Itti, L., & Munoz, D. P. (2012). Linking visual response properties in the superior colliculus to saccade behavior. *European Journal of Neuroscience, 35*(11), 1738–1752.

Masri, R. A., Percival, K. A., Koizumi, A., Martin, P. R., & Grünert, U. (2016). Survey of retinal ganglion cell morphology in marmoset. *The Journal of Comparative Neurology, 524*(9), 1839–1858.

Matin, E. (1974). Saccadic suppression: A review and an analysis. *Psychological Bulletin, 81*(12), 899–917.

Maunsell, J., Nealey, T., & DePriest, D. (1990). Magnocellular and parvocellular contributions to responses in the middle temporal visual area (MT) of the macaque monkey. *The Journal of Neuroscience, 10*(10), 3323–3334.

May, P. (2014). Superior Colliculus. In *Encyclopedia of the Neurological Sciences* (Vol. 1, pp. 351–354). San Diego, CA: Academic Press.

Merigan, W. H., & Maunsell, J. H. R. (1993). How parallel are the primate visual pathways? *Annual Review of Neuroscience, 16*(1), 369–402.

Milner, A. D., & Goodale, M. A. (2007). Two visual systems re-viewed. *Neuropsychologia, 46*, 774–785.

Munoz, D. P., & Fecteau, J. H. (2002). Vying for dominance: Dynamic interactions control visual fixation and saccadic initiation in the superior colliculus. *Progress in Brain Research, 140*, 3–19.

Munoz, D. P., & Schall, J. D. (2004). Concurrent, distributed control of saccade initiation in the frontal eye field and superior colliculus. In W. C. Hall & A. Moschovakis (Eds), *The Superior Colliculus: New Approaches for Studying Sensorimotor Integration* (pp. 55–82). New York: CRC Press.

Munoz, D. P., & Wurtz, R. H. (1993). Fixation cells in monkey superior colliculus. I. Characteristics of cell discharge. *Journal of Neurophysiology, 70*(2), 559–575.

Nassi, J. J., & Callaway, E. M. (2009). Parallel processing strategies of the primate visual system. *Nature Reviews Neuroscience, 10*(5), 360–372.

Nassi, J. J., Lyon, D. C., & Callaway, E. M. (2006). The parvocellular LGN provides a robust disynaptic input to the visual motion area MT. *Neuron, 50*(2), 319–327.

Nealey, T. A., & Maunsell, J. H. (1994). Magnocellular and parvocellular contributions to the responses of neurons in macaque striate cortex. *The Journal of Neuroscience, 14*(4), 2069–2079.

Nguyen, M. N., Matsumoto, J., Hori, E., Maior, R. S., Tomaz, C., Tran, A. H., Ono, T., & Nishijo, H. (2014). Neuronal responses to face-like and facial stimuli in the monkey superior colliculus. *Frontiers in Behavioral Neuroscience, 8*, 85.

O'Regan, J. K., & Noë, A. (2001). A sensorimotor account of vision and visual consciousness. *Behavioral and Brain Sciences, 24*(05), 939–973.

Olivier, E., Dorris, M. C., & Munoz, D. P. (1999). Lateral interactions in the superior colliculus, not an extended fixation zone, can account for the remote distractor effect. *Behavioral and Brain Sciences, 22*(4), 694–695.

Palmer, S. E., & Brooks, J. L. (2008). Edge-region grouping in figure-ground organization and depth perception. *Journal of Experimental Psychology: Human Perception and Performance, 34*(6), 1353–1371.

Petersen, S. E., & Posner, M. I. (2012). The attention system of the human brain; 20 years. *Annual Review of Neuroscience, 35*(1), 73–89.

Peterson, M. A., & Cacciamani, L. (2013). Toward a dynamic view of object perception. In S. J. Dickinson & Z. Pizlo (Eds), *Shape Perception in Human and Computer Vision* (pp. 443–457). London: Springer-Verlag.

Peterson, M. A., de Gelder, B., Rapcsak, S. Z., Gerhardstein, P. C., & Bachoud-Lévi, A. C. (2000). Object memory effects on figure assignment: Conscious object recognition is not necessary or sufficient. *Vision Research, 40*(10), 1549–1567.

Previc, F. H. (1990). Functional specialization in the lower and upper visual fields in humans: Its ecological origins and neurophysiological implications. *Behavioral and Brain Sciences, 13*(3), 519–575.

Purves, D., Morgenstern, Y., & Wojtach, W. T. (2015). Perception and reality: Why a wholly empirical paradigm is needed to understand vision. *Frontiers in Systems Neuroscience, 9*, 156.

Rayner, K. (1998). Eye movements in reading and information processing: 20 years of research. *Psychological Bulletin, 124*(3), 372–422.

Sakaguchi, Y. (2001). Target/surround asymmetry in perceptual filling-in. *Vision Research, 41*(16), 2065–2077.

Schiller, P. H., & Tehovnik, E. J. (2009). Cortical control of eye movements. In L. R. Squire (Ed.), *Encyclopedia of Neuroscience* (Vol. 3, pp. 175–181). Oxford: Academic Press.

Sincich, L. C., & Horton, J. C. (2005). The circuitry of V1 and V2: Integration of color, form, and motion. *Annual Review of Neuroscience, 28*, 303–326.

Sun, H., Smithson, H. E., Zaidi, Q., & Lee, B. B. (2006a). Specificity of cone inputs to macaque retinal ganglion cells. *Journal of Neurophysiology, 95*(2), 837–849.

Sun, H., Smithson, H. E., Zaidi, Q., & Lee, B. B. (2006b). Do magnocellular and parvocellular ganglion cells avoid short-wavelength cone input? *Visual Neuroscience, 23*(3–4), 441–446.

Tolhurst, D. J. (1975). Sustained and transient channels in human vision. *Vision Research, 15*, 1151–1155.

Troxler, I. P. V. (1804). Über das Verschwinden gegebener Gegenstände innerhalb unseres Gesichtskreises [Disappearance of objects presented within our field of vision]. In K. Himly & J. A. Schmidt (Eds), *Verschwinden Unseres Ophthalmologischen Bibliothek* (Vol. 2, pp. 275–278). Jena: Fromann.

Tzelepi, A., Laskaris, N., Amditis, A., & Kapoula, Z. (2010). Cortical activity preceding vertical saccades: A MEG study. *Brain Research, 1321*, 105–116.

Ungerleider, L. G., & Haxby, J. V. (1994). "What" and "where" in the human brain. *Current Opinion in Neurobiology, 4*(2), 157–165.

Ungerleider, L. G., & Mishkin, M. (1982). Two cortical visual streams. In D. J. Ingle, M. A. Goodale, & R. J. W. Mansfield (Eds), *Analysis of Visual Behavior* (pp. 549–586). Cambridge, MA: MIT Press.

Vecera, S. P. (2000). Toward a biased competition account of object-based segregation and attention. *Brain and Mind, 1*(3), 353–384.

von der Heydt, R. (2015). Figure-ground organization and the emergence of proto-objects in the visual cortex. *Frontiers in Psychology, 6*.

Weisstein, N. (1972). Metacontrast. *Handbook of Sensory Physiology*, 7(4), 233–244.

Weisstein, N., & Brannan, J. R. (1991). A low spatial frequency, red sine wave grating will float in front of gratings with the same or similar spatial frequency but other chromaticities: M and P interactions in figure-ground perception. *Investigative Ophthalmology & Visual Science, 32*, 1274.

Weisstein, N., & Wong, E. (1987). Figure-ground organization affects the early visual processing. In M. A. Arbib & A. R. E. Hansen (Eds), *Vision, Brain, and Cooperative Computation* (pp. 209–230). Cambridge, MA: MIT Press.

Weisstein, N., Maguire, W., & Brannan, J. R. (1992). M and P Pathways and the perception of figure and ground. In J. R. Brannan (Ed.), *Applications of Parallel Processing in Vision* (pp. 137–166). New York: Elsevier Science Publishers B.V.

White, B. J., & Munoz, D. P. (2011). Separate visual signals for saccade initiation during target selection in the primate superior colliculus. *The Journal of Neuroscience, 31*(5), 1570–1578.

Wiesel, T. N., & Hubel, D. H. (1966). Spatial and chromatic interactions in the lateral geniculate body of the rhesus monkey. *Journal of Neurophysiology, 29*(6), 1115–1156.

Williams, M. C., & Weisstein, N. (1981). Spatial frequency response and perceived depth in the time-course of object superiority. *Vision Research, 21*(5), 631–646.

Wokke, M. E., Scholte, H. S., & Lamme, V. A. F. (2014). Opposing dorsal/ventral stream dynamics during figure-ground segregation. *Journal of Cognitive Neuroscience, 26*(2), 365–379.

Wokke, M. E., Sligte, I. G., Scholte, H. S., & Lamme, V. A. F. (2012). Two critical periods in early visual cortex during figure–ground segregation. *Brain and Behavior, 2*(6), 763–777.

Wong, E., & Weisstein, N. (1982). A new perceptual context-superiority effect: Line segments are more visible against a figure than against a ground. *Science, 218*(4572), 587.

Wong, E., & Weisstein, N. (1983). Sharp targets are detected better against a figure, and blurred targets are detected better against a background. *Journal of Experimental Psychology: Human Perception & Performance, 9*(2), 194–201.

Wong, E., & Weisstein, N. (1984). Flicker induces depth: Spatial and temporal factors in the perceptual segregation of flickering and nonflickering regions in depth. *Perception & Psychophysics, 35*(3), 229–236.

Wong, E., & Weisstein, N. (1985). A new visual illusion: Flickering fields are localized in a depth plane behind nonflickering fields. *Perception, 14*(1), 13–17.

Wong, E., & Weisstein, N. (1987). The effects of flicker on the perception of figure and ground. *Perception & Psychophysics, 41*(5), 440–448.

Wurtz, R. H. (2008). Neuronal mechanisms of visual stability. *Vision Research, 48*(20), 2070–2089.

2

PHANTOMS AT THE HOLIDAY INN

Patrick Cavanagh
Glendon College

ARVO is a vision conference, now superseded by VSS (the Vision Sciences Society), but like VSS it was originally held in Florida at a beach front hotel. The current ARVO is a staid affair, having drifted away from Florida and its original mission, and now attended mostly by ophthalmologists, but in its early era it was an interdisciplinary hot bed of vision research, wildly successful undoubtedly because of its location on beautiful white sand beaches. In 1977, it was the first vision conference I attended and I was doing my best to keep up and make sense of the flood of new ideas and excitement. The meeting room was a Holiday Inn conference venue perched on the Lido Key beach facing the Gulf of Mexico. Much of our day, at least noon to 4 pm was spent on that beach talking and playing but now, later in the day, we were on the top floor of the hotel and the room was packed to hear Naomi Weisstein. This was not science as I knew it. Her talk was punchy, full of jokes and new ideas about visual cognition, phantoms, and a demonstration. A show-stopping demonstration. Off went the room lights and up came a movie (no videos at that time) of vertical strips of Xs interrupted in the middle by a dark horizontal band (Figure 2.1). "Look at the way the light and dark bars complete across the strip," she said, and there was a murmur of agreement. Magical phantoms crossed the gap, which I learned later had been previously reported (Tynan & Sekuler, 1975). But forget that, Naomi was telling us that not only were there phantoms, but they also generated motion aftereffects in the gap where nothing had been present. These findings were reported in December of that year in a *Science* paper (Weisstein et al., 1977).

On the left is an original frame digitized from the 1977 demonstration movie with arrows overlaid to indicate the motion (see Movie 1 in the list at end of chapter). In the center is a close up depiction of what the authors, Weisstein, Maguire, and Berbaum (1977) claimed they and their observers saw – light and

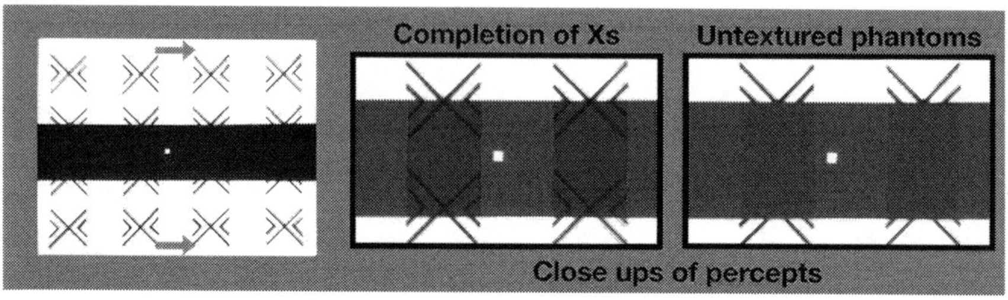

FIGURE 2.1 Phantom texture completion.

dark bars completing across the gap in the same phase as the moving stripes but importantly with a faint continuation of the X pattern. The contrast of the induced patterns is exaggerated here. On the right is a close up of what most would agree is visible when the grating moves – the phantom stripes but without any texture completion. The contrast is again exaggerated compared to what observers actually see. This chapter chronicles my efforts to track down the elusive extrapolating Xs.

Back in the presentation hall, oohs and aahhs started gradually and then were overwhelmed by another commotion as she said that some of us should also see the Xs complete through the gap. I believe I remember seeing them. Clearly some others did as well, because several called out for her to stop the projector on a single frame to prove there was no trickery. "Stop the projector!" The projectionist stopped the movie on a single frame – there were no Xs in the gap. "Start it again!" The movie clattered into action and there were the Xs again, at least for some of us. Enough to get a good measure of applause.

That was my introduction to the immediacy and excitement of the new vision science. It was a special moment when physiology and computer vision and psychophysics were joining together to find new questions, answers and models. Fourier theories were in the air. Crackling presentations by Bela Julesz, Stuart Anstis, David Marr, John Robson, Richard Gregory, and Naomi made us feel like the best and smartest were converging on vision science as the gateway to a new understanding of the brain. We were getting high on science. Naomi herself was bursting with charisma and confidence and intelligence. She chortled with each point she made, "hah", like the fist pumps that tennis players make when they baffle an opponent. But sports players are rather humorless (except for Spaceman Bill Lee, the legendary pitcher of the Montreal Expos) whereas Naomi had humor in abundance.

As you can judge from my retelling, Naomi's talk was imprinted on my new vision scientist's mind as an astounding finding opening new directions and I have spent numerous moments in the last 40 years tracking it down. In their *Science* paper they reported the extrapolating Xs in only one sentence and spent the rest of the article describing the motion aftereffect they measured on a test placed in the unstimulated gap. Even so, they were very confident of the effect. Naomi even told Bob Sekuler that not just Xs would complete into the gaps. If she had put birds in the columns instead of Xs, she said, we would not only see birds in the gap but we could also report what kind of bird it was.

However, in our ordinary labs, it was not evident that any Xs, let alone birds, would ever appear in the staunchly black occluding stripe. Yes, we could see the phantom stripes of light and dark faintly crossing the gap but they were quite dim and always untextured. Our first attempts with Peter Tse, 15 years after the ARVO presentation, used regular computer controlled video. This failed to generate any texture completion and we thought it might be the differences in the properties of the monitors compared to the movie. Or perhaps we had all been simply in the thrall of Naomi's wonderful presentation skills and had undergone a mass induction. Naomi and her colleagues and Sekuler published other papers on this moving phantom grating effect (Brown & Weisstein, 1988, 1991; Brown, 1993; Mulvanny et al., 1982) but there were no further tests of the texture completion with moving gratings. In Japan, Akiyoshi Kitaoka, Jiro Gyoba, Hideaki Kawabata, and Kenzo Sakurai explored many versions of static gratings that produced phantoms but without any texture completion (until recently, see their chapter in this book). What was up? Was there nothing here? But first, why was Naomi's claim of pattern extrapolation interesting anyway?

When a pattern is partially hidden, we do represent the missing parts behind the occluding surface, or at least our brain does, somewhere. When we see a dog's head sticking up behind a garden gate, for example, we do not take it as an unattached head; instead, our visual system represents the missing body. But importantly we do not see or hallucinate the missing body as if it were visible through the door—it has a more abstract, non-visual representation. The phantom grating, originally reported by Tynan and Sekuler (1975), is an exception. If the occluded pattern is moving and present on both sides of the occluder, it is no longer just represented notionally as going behind the gap but completes in front as if it were a dim transparency that is sliding over the occluder. Perhaps it is just easier to have some visible token that is moving rather than continually updating an invisible, hidden token. We really do not know why this happens, but the important part, at least to start, is that the phantom is visible and it gives an observable percept to help explore and understand these completion processes.

Naomi made the completion do work, by showing a motion aftereffect in the physically unstimulated central portion. But we want to do more. What patterns would complete? How detailed would they be? Would fMRI show activation to the patterns in the gap? If different patterns or colors were on opposite sides of the gap, would there be a transition from one to the other cross the gap as happens in apparent motion between dissimilar stimuli? Kitaoka, Gyoba, and Sakurai (2006) showed a related chimeric completion of static, fuzzy, untextured vertical bars across a grey occluder when the top background was black and the bottom was a random dot pattern. If the vertical bars had texture elements would they too bridge the gap? If they were, say, Xs on top and Os on the bottom, what would the intermediate phantom texture elements look like? Answers to any of these questions would help understand how vision fills in and completes patterns from partial input and where this critical step happens in the various stages of processing in the visual system.

This filling in is of course not the only example we have of completion. Perhaps the most famous is completion across the blind spot where textures and lines and colors complete across the gap where there is no sensory input (see review Ramachandran, 1992). Similar reports are found for other losses of sensory input, scotomas, caused by retinal or cortical damage. However, the nature of the completion, at least for the blind spot, is problematic. Everyone agrees that they see no gap in the blind spot but because it is in peripheral vision it is hard to say whether anything is actually visible there. It might be just a case of thinking the pattern must be filled in because there is no signal to say it is not. Whatever your opinion on this philosophical issue, the practical problem is that the area of the visual field translates to only a small portion of cortex so it is hard to use fMRI to examine what representation there is at that location.

And this is the importance of Naomi's phantom textures. They are right in the fovea and should fill substantial cortical real estate that would offer lots of signal in fMRI. Here again though, there has not been much success. Meng, Remus, and Tong (2005) were able to get phantom contours to cross a gap in the middle of a drifting sinewave grating and they could decode the presence of the induced activation in the cortical areas that respond to the gap where there was no physical stimulus. But they did not find any texture completion across the gap (personal communication, Ming Meng, 2017).

Perhaps the lack of success in all the follow-up attempts was due to the difference between video and movies. To check this, I wanted to duplicate Naomi's projection version and began an effort to recreate the 1977 event. In 2004, I asked Naomi to send me the actual 16 mm film version she had used. She promptly did. I am not sure how she had uncovered it in the various boxes or files that she had kept over the years, but here it was in my hands. Unfortunately, I couldn't get a 16 mm projector through the audio visual services. They no longer had any. Not unreasonable considering that no one had used 16 mm films in classes for maybe 20 years. I spent the next 10 years pursuing other projects. I moved to Dartmouth in 2015 where again I could not obtain any projector from the audio visual services but finally I bid on one on eBay and won it. It arrived in the summer of 2016 and we quickly tried a screening of the movie. There was a lot of mechanical clattering and the jumping film spewed onto the floor before we learned how to properly thread and maintain the forward progress of the film. I think it is a good thing that we no longer rely on 16 mm projectors. But again, disappointment. We saw some grating completion but no convincing Xs in the gap and I worried that maybe Naomi had used a better projector or a different viewing distance. My newly acquired 1960s Bell and Howell model had unstable speed and only a 1 blade shutter giving 24 Hz and a 50 percent duty cycle. High end projectors had 3 blade shutters and a flicker rate of 72 Hz.

Without Naomi to goad us on with encouragement and guffaws, the momentum slowed. But did not fade away. Gyoba, Sakurai, and Kitaoka tried further moving versions (see their chapter here) and presented them at conferences. Observers reported something vaguely textured in the gap.

We generated new computer versions, some that even matched the flicker quality of Naomi's movie and distributed them to several hard-core vision scientists for evaluation (see Movie 3 at end of chapter). Many said they saw nothing but some admitted they could be seeing some texture in the gap, not necessarily visible texture but a feeling that something might be there as if a back texture were sliding over a black surface. This was a long way from the original effect which caused such a stir and even farther from the completion of identifiable birds in the gap. So we tried a few last variations.

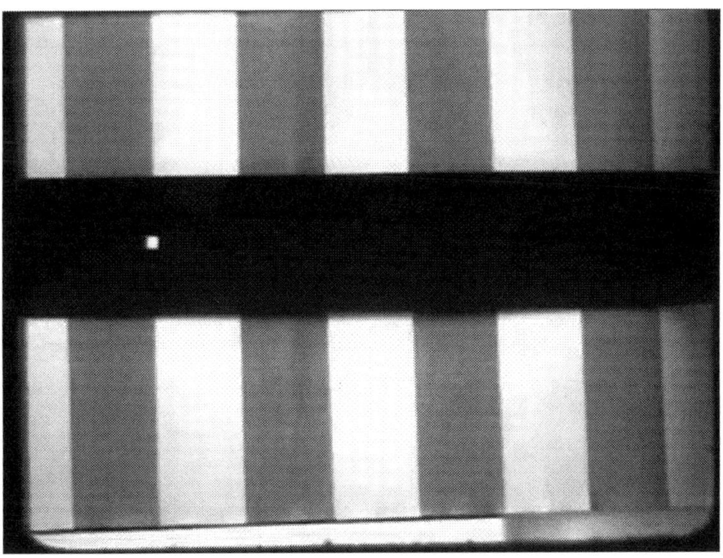

FIGURE 2.2 The scroll of paper with a light and dark grating on it that moved like a conveyor belt around two vertical spindles.

This was the apparatus in the SUNY Buffalo lab that Naomi and her colleagues used to test the phantom gratings, here for simple induction without texture. The dark horizontal strip was a piece of black board spanning the moving display, with a small white fixation mark that was centered in the full display. This version can be seen in Movie 2 (see movie list at the end of the chapter) and produces strong phantom gratings.

The flicker aspect of the movie version did not turn out to be important and that seems reasonable given that the original tests were run on a set of textured stripes printed onto a scroll of paper that was wound around two drums. Unlike the movie, it was seen in steady light and drifted continuously, powered by a little motor very much like the original devices used by Munsterberg and others to study motion aftereffects at the end of the 1800s and beginning of the 1900s. The apparatus is not described in the article but we can see it in the brief docudrama at the end of Naomi's demonstration movie (Movie 4 below) which records an apparently inadvertent release of the phantoms from the grating. They invade the lab, driving the authors out to the street. We can also see a slight flap at the incompletely glued joint between the two ends of the printed scroll. A horizontal strip of cardboard was positioned in the middle of the grating to create the occluding gap.

Our final tests for this chapter with the movie and the video versions were at a greater viewing distance to mimic the visual angle of the 1977 audience at the back of the hall. Indeed, one skeptical and steely eyed observer felt that the texture was visible spilling across the edges of the occluder but not all the way into its center. The greater viewing distance would narrow the gap in terms of visual angle and perhaps maximize the encroachment of the texture into the gap. The partial completions are the best we have to report and Naomi's original displays were constructed so that the X textures only needed to extend into the gap at the edges (see Figure 2.1 above) – the spacing left only a gap in the middle of the occluder, without room for a full phantom X in the center.

Naomi's charisma and her stream of novel discoveries – object superiority, phantom gratings, foreground biases, and others – filled a remarkable six *Science* papers (and many others) in the brief span before her illness halted her work. This work brought lots of attention to the new field of visual cognition and we are indebted to her pioneering work in many ways. Many of us who appreciated and followed up on her science were not aware that she was even more famous in two other fields – stand-up comedy and the feminist movement, where she led a feminist rock band on tracks like "Papa don't lay that shit on me". Her feminist writings were so influential that Duke University paid a seven figure amount to get the full catalog of her work, and MS magazine ran a feature article on her. Even if we were unaware of her other selves, we were in no way surprised by her successes in these areas given the drive, wit and creativity that we were so lucky to witness in her scientific presentations.

Movies

To view movies for Chapter 2 go to https://research.franklin.uga.edu/visual-perception-laboratory/chapter-2

Movie 1 Naomi's original movie. For best results, view at a dim brightness setting in a dark room with nothing else on the monitor (no bright background). Find a viewing location where the image subtends around 5° to 10° of visual angle. Make sure it loops.

Movie 2 Here is the simple light/dark moving grating from the original demonstration movie. I find it produces the best induction of phantom stripes into the gap. I have to block off the grating to convince myself there are not artifacts in the gap. Make sure it loops.

Movie 3 This is a computer generated version of Naomi's movie. Again, for best results, view at dim brightness in a dark room with nothing else on the monitor. You may see some completion of the light and dark stripes across the gap. Some may see some intrusion of the texture into the gap. There is no fixation spot, please add one with the mouse cursor if that helps.

Movie 4 At the end of the 16 mm demonstration film was a quirky, sci-fi docudrama relating what happened in Naomi's lab when the phantoms escaped from the apparatus. Very much in the Beatles "Hard Day's Night" style. The two others in the film are Bill Maguire and Kevin Berbaum, her co-authors of the 1977 paper.

Acknowledgments

The work reported here was funded by grants in Canada (NSERC), the US (NIH, AFOSR, Harvard and Dartmouth Startup funds), and France (ANR, ERC) across a span of 40 years. Many thanks to Bill Maguire, Kevin Berbaum, Christopher Tyler, Bob Sekuler, Ming Meng and others for comments and observations. And thanks to Jesse Lemisch and Naomi Weisstein for the thrill of the quest.

References

Brown, J. M. (1993). Moving phantom visibility as a function of fundus pigmentation. *Perception & Psychophysics, 3*, 367–371.

Brown, J. M., & Weisstein, N. (1988). A phantom context effect: Visual phantoms enhance target visibility. *Perception & Psychophysics, 43*, 53–56.

Brown, J. M., & Weisstein, N. (1991). Conflicting figure-ground and depth information reduces moving phantom visibility. *Perception, 20*, 155–165.

Kitaoka, A., Gyoba, J., & Sakurai, K. (2006). The visual phantom illusion: A perceptual product of surface completion depending on brightness and contrast. In S. Martinez-Conde, S. L. Macknik, L. M. Martinez, J-M Alonso, & P. U. Tse (Eds). *Progress in Brain Research* (Vol. 154, pp. 247–262). Amsterdam: Elsevier.

Meng, M., Remus, D. A., & Tong, F. (2005). Filling-in of visual phantoms in the human brain. *Nature Neuroscience*, *9*, 1248–1254.

Mulvanny, P., Macarthur, R., & Sekuler, R. (1982). Threshold for seeing visual phantoms and moving gratings. *Perception*, *11*, 35–46.

Ramachandran, V. S. (1992). Blind spots. *Scientific American*, *266*(5), 86–91.

Tynan, P., & Sekuler, R. (1975). Moving visual phantoms: A new contour completion effect. *Science*, *188*, 951–952.

Weisstein, N., Maguire, W., & Berbaum, K. (1977). A phantom-motion aftereffect. *Science*, *198*, 955–958.

3

OBJECT-SUPERIORITY AND PHANTOM GRATINGS

Top-Down Influences in Organization of Bottom-Up Visual Processing

Bruno G. Breitmeyer
University of Houston

A central issue pondered by philosophers through the ages is the relation between the One and the Many. It is an overarching issue that implicitly, if not explicitly, crops up in most scholarly and scientific enterprises. Is the One simply the sum of the Many, or is it something more than their sum? Since perception entails the phenomenal, subjective experience of the world in which each of us exist, the issue can be posed as to whether each of our conscious visual fields is unitary over and above and despite its manifold contents or, in contrast, whether it simply is nothing but the sum of its contents. In the last century and a half of research in visual perception this global issue has motivated perennial research programs that focus more modestly on exploring and uncovering the relation between a holistically apprehended visual object, the one, and its many parts. The research programs have been clothed in various fashionable attire: structuralism, looking at the mind in terms of mental chemistry, with its key assumptions phrased in terms of the constancy hypothesis (sensory atoms) and the bundle hypotheses (the rules of bonding or associating sensory primitives to form wholes (mental molecules)); Gestalt psychology with its emphasis on non-derived wholes and on the holistic organizational principles governing their parts, all framed within a field-theoretic approach to phenomenal experience and its isomorphic neural organization of the brain; the variety of behaviorist approaches that, with a few exceptions, by and large eschewed any study of the "black box" intervening between sensory input and motoric output; the information-processing approach of cognitive psychology that reveled in creating a slew of black boxes and the communication of information among them; and more recently, with advances in the neurosciences, the neo-structuralist—or, if you prefer, the neo-Gestalt—approach that recognizes the importance of holistic organization in visual perception yet also looks to see how the visual analysis of underlying parts and their relations contribute to that organization and conversely how the holistic organization relates to the analysis of parts.

In the late 1950s, under the influence of cybernetic (Wiener, 1948), computing (Turing, 1950; von Neumann, 1958), and communication (Shannon & Weaver, 1949) principles, the notion of feedback in information processing became a central component in cognitive psychology. In its fledgling versions, it was expressed in the Test-Operate-Test-Exit (TOTE) concept (Miller et al., 1960), and in the notion of analysis-by-synthesis (Neisser, 1967). The central idea relevant to information processing was that the existence of holistic representation, be it perceptual, linguistic or cognitive, interacted via feedback with its components that in their relation to each other constitute it. Limiting ourselves to visual perception, application of the basically cybernetic TOTE or analysis-by-synthesis approach rests on several key notions: 1) the generation of "set points", i.e., tentative holistic representations that can be considered as an analog of "perceptual hypotheses", which via feedforward-feedback loops are tested against incoming information; 2) the generation of corresponding error signals, of which the smallest "confirms" its corresponding hypothesis and larger ones "disconfirm" the other corresponding perceptual hypotheses. Moreover, 3) should this iterative process fail due to a persistently large error signal, either new perceptual hypotheses are generated, i.e., new set points are acquired and/or 4) more information is acquired and Steps 1 and 2 are repeated until all but one of the perceptual hypotheses is eliminated. Fortunately, *almost* always, the first two steps are sufficient in our daily commerce with the visual world around us. However, most of us can also remember the rare instances of our having to perform a double take before an initial nonveridal percept is corrected. Like the concepts of the One and the Many, implicitly, if not explicitly, these notions were floating around and influenced a variety of cognitive-psychology paradigms in the 1960s and 1970s, including of course research on visual object and pattern recognition.

Within this paradigmatic context, Naomi Weisstein's research conducted in the 1970s and 1980s made several important pioneering contributions. Although during those years and earlier she conducted significant work on visual masking (Weisstein, 1968; Weisstein et al., 1975; see Öğmen's contribution, this volume) and foreground-background or figure-ground segregation (Wong & Weisstein, 1982, 1987; see Gillam's, Peterson's, and Brown & Greene's contributions, this volume), I will focus on two other visual phenomena: the object-superiority effect (Weisstein & Harris, 1974) and phantom gratings (Weisstein et al., 1977; see Cavanagh's and Gyoba, Sakurai & Kitaoka's contributions, this volume). These two phenomena, like the others she worked on, are very germane to recent and ongoing research in visual cognition and neuroscience.

Object-Superiority Effect

Visual perception can be categorized according to a hierarchy of complexity starting with simple, primitive 1-D features such as the orientations of lines or their colors, which when conjoined can give rise to the percepts of 2-D structures and shapes (e,g., red rectangles, blue triangles, etc.), which in turn, when further conjoined, can yield percepts of higher-order structures of 3-D objects (e.g., Rubik's cubes). The question posed by Weisstein and Harris (1974) was whether a holistic 3-D object representation can affect the perception of its parts. Shortly before Weisstein and Harris (1974) investigated the object-superiority effect, its lexical analogue, the word-superiority effect, had already been reported by Reicher (1969) and Wheeler (1970). Figure 3.1 illustrates stimuli used to study the latter effect. Here for example, a letter R or S embedded in a word CAST or CART, which is flashed briefly and is immediately followed by a powerful pattern mask, is less subject to be masked than the same letter when it is embedded in a nonword TCSA or ATRC. That is to say, a mask that is just powerful enough to produce, say, chance (50 percent correct) performance in correctly identifying the (third) test letter in either of the two nonwords will yield performance significantly above chance when the same test letter is embedded in either of the two words. Apparently, relative to when it is embedded in nonwords, its presence in the word "immunizes" its visibility to the powerful masking effect. However, subsequent work by Baron and Thurston (1973) showed i) that embedding a target letter in a word is not a necessary condition for obtaining a superiority effect, since embedding it in pronounceable nonwords such as VUST or VURT, as compared to much less pronounceable nonwords such as UTSV or UTRV, produce a

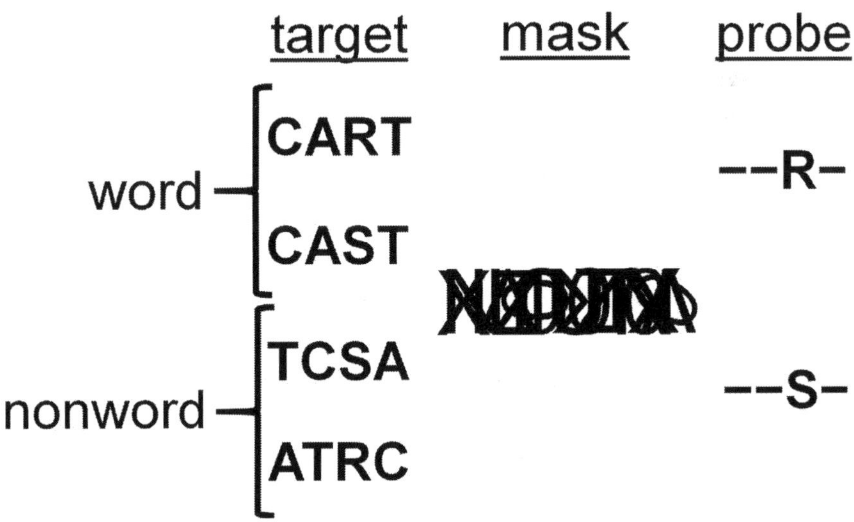

FIGURE 3.1 Examples of stimuli used to study the word-superiority effect.

A target letter, here R or S, is embedded in a word, CART or CAST, or in a nonword, ATRC or TCSA. These word or nonword letter sequences are presented briefly and are followed by a patterns mask. The visibility of the masked R or S when embedded in a word is higher than when embedded in a nonword.

superiority effect equal to that of embedding it in words; and ii) that the superiority effect is related more to orthographic or spelling regularities than to meaningful (word) or phonological (sound) associations of letter sequences in which the target is embedded. So, it is not words per se, but orthographic regularities supervening on individual letters, that are responsible for the word-superiority effect.

Weisstein and Harris (1974) wondered if, similarly, a holistic superiority also held for nonlinguistic and abstract visual shapes supervening on their components. They created context stimuli (for examples see Figure 3.2) that varied in their perceived object quality when they were conjoined with the target stimuli. The target-plus-context stimuli could appear vividly like 3-D objects yielding high subjective depth ratings (left display in lower panel), or less so and with subjective depth ratings tending more toward flat or 2-D shapes in which the line segments comprising the target and the context are somewhat connected (middle display in lower panel) or totally disconnected (right display in lower panel) from

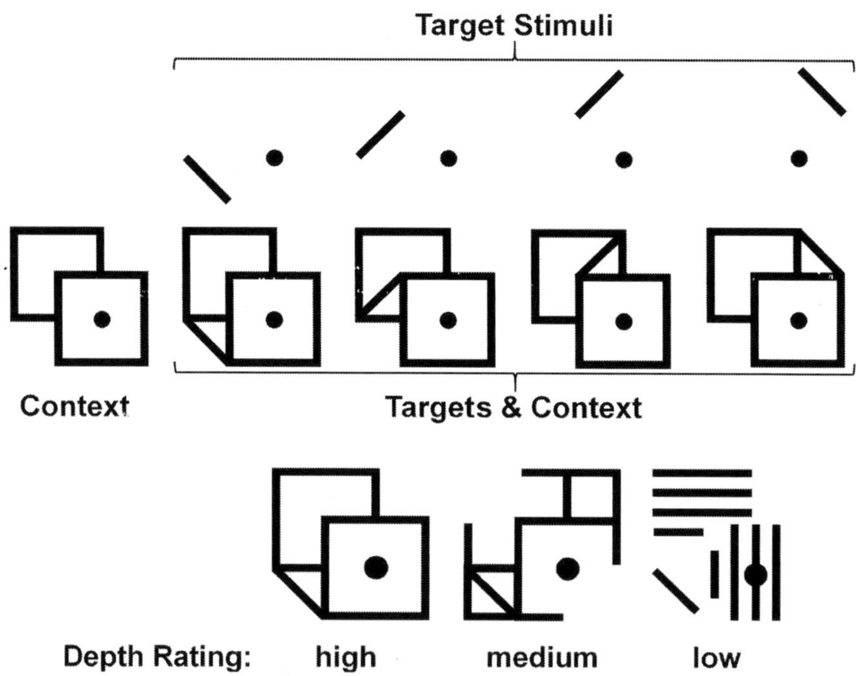

FIGURE 3.2

Upper panels: Depictions of target lines along with a context in which they are embedded that, in this example, yield high ratings of object-depth ratings. Lower panel: examples of a target line embedded in three contexts yielding successively lower object-depth ratings. The targets embedded in their contests are presented briefly and are immediately followed by a powerful pattern mask. When masked, target lines embedded in displays yielding higher object-depth ratings are more visible.

each other. The targets were presented briefly and were followed immediately by a powerful pattern mask. The results showed that, as the object-like perceptual quality of the target-plus-context displays progressively decreased from a 3-D representation to the least object-like level of a 2-D representation of disconnected line segments, identification performance of the targets, measured in percentage correct, decreased by nearly 20 percent. The upshot, as in the word-superiority effect, is that the visibility of targets embedded, here, in 3-D object-like stimuli are also immunized against an otherwise powerful masking effect. This indicates that holistically perceived visual objects can effectively increase the visibility of their parts. McClelland and Miller (1979) argued, however, that it is the structural relevance of the target within its context that determines its superior visibility rather than its embeddedness in a 3-D object representation. In a follow-up study, in which ratings of structural relevance correlated strongly ($r = 0.98$) with depth ratings of the contexts, Weisstein, Williams and C. S. Harris (1982b) found that the object-superiority effect decreased as the structural relevance as well as 3-D appearance of the context decreased. Because of the strong correlation between structural relevance and 3-D ratings, the experiment could not adjudicate which of the two factors was responsible for the object-superiority effect. However, in a second follow-up study, Lance, Weisstein and J. R. Harris (1982) were able to construct context stimuli in which observers' depth ratings of these stimuli did not correlate with their structural-relevance ratings ($r = 0.02$) and, more importantly, in which the depth ratings correlated positively with target visibility ($r = 0.81$) but did not correlate with structural-relevance ratings ($r = -0.05$). These findings indicated that it is primarily the perceived three dimensionality of the context stimuli and not their structural relevance that gives rise to the object-superiority effect. Subsequent studies, using a variety of psychophysical methodologies, were undertaken by Noami Weisstein and her students that followed up on and solidified these basic findings; however, they will not be reviewed here.

Phantom Gratings

Like many other illusory percepts that have become fashionable in recent vision research, phantom percepts were investigated by Gestalt psychologists as early as the beginning of the twentieth century (Rosenbach, 1902), and were reintroduced to vision research in the 1970s by Tynan and Sekuler (1975) using the then fashionable sinusoidal gratings as inducing stimuli. Figure 3.3 shows an example of the induction of stationary phantom gratings in a region in which gratings are not physically present. When viewing a display like that shown in Figure 3.3, many observers report seeing the dark sinusoids extending through the black rectangle. The perception of illusory gratings becomes strikingly vivid when the grating drifts horizontally or flickers (Maguire & Brown, 1987; Genter & Weisstein, 1981). Moreover, from a neural-activation standpoint, the moving

FIGURE 3.3 A typical phantom grating display.

illusory gratings are sufficiently "real" to produce motion after-effects in the blank region in which they were perceived (Weisstein et al., 1977; Weisstein et al., 1982a). Again, follow-up studies undertaken by Noami Weisstein and her students corroborated these basic findings and their theoretical interpretations, some of these Weisstein-inspired studies extending into the current century (e.g., Brown et al., 2001).

Assessments of Current Implications and Impact

However, the legacy of these influential studies extends still further, into significant areas of recent research in visual cognition and neuroscience (Fox et al., 2017; Herzog & Clarke, 2014; Komatsu, 2006; Pomerantz & Cragin, 2015; Pomerantz & Portillo, 2011), particularly in the areas of configural perceptual organization, visual object recognition/perception and the roles of filling-in processes in visual perception. To lay the groundwork for the following discussion, a brief theoretical overview of what is involved in object recognition and/vs. object perception is warranted.

Object-superiority and phantom gratings are visual effects that give rise to our perceptual experience of an object's depth and of illusory presences that are lacking in the purely physical description of the stimuli giving rise to them. Nonetheless, our phenomenal experience, though subjective, is amenable to objective assessments of various kinds. Object and pattern recognition can of course proceed without phenomenal experience. It can be performed by artificial-intelligence devices programmed to, for instance, accomplish simple tasks like recognizing the bank-routing and personal-account numbers on our checks or, as in the more complex case of driverless cars, the objects to avoid and spaces through which to proceed in traffic environments. However, object perception, i.e. their phenomenal experience is – at least up to now – uniquely tied to sentient creatures. This phenomenological aspect, so important to many perceptual effects besides object superiority and phantom gratings, is, as noted by Herzog, Sayim, Chicherov and Manassi (2015), not subject to a "naturalizing" via reductive mathematical modelling that can be instantiated in computational devices.

Though not reducible to computational modelling, the properties of phenomenal object perception nevertheless can be related to neural models of visual perception, a particularly relevant example of which is the neuro-computational approach developed by Grossberg and co-workers (Cao & Grossberg, 2005; Grossberg, 1994; Grossberg & Yazdanbakhsh, 2005). These models rest on the fundamental distinction between the shape and surface properties of visual objects. The simplest version distinguishes correspondingly between a neural boundary-contour-system (BCS), which at preconscious levels computes the shape properties defined by the boundaries or contours of an object, which are filled in by its surface properties derived from the neural feature-contour-system (FCS). It is only after or as the filling-in process occurs that the object attains its phenomenally perceived state, shape and all (Grossberg, 2003). Although this theoretical distinction between shape and surface attributes of visual objects is supported from a phenomenological standpoint, is it also supported by what we can glimpse of the functional architecture of the visual system? A review of the neuroanatomy and neurophysiology of the primate visual system indicates that it is quite realistic and has particularly important implications for filling-in processes beyond those applying to object surfaces.

Neuronal Correlates of the BCS and FCS

Evidence accumulated over the past two decades indicates that there are anatomically identifiable pathways and areas in the early and intermediate cortical object-processing systems that process primarily the surface properties of color and lightness on the one hand and the form properties of contour and edge orientation on the other (Conway et al., 2007; Felleman et al., 1997; Kinoshita & Komatsu, 2001; Lu & Roe, 2008; Roe & Ts'o; 1999; Sincich & Horton, 2005; Wang et al., 2007; Xiao, Casti, Xiao & Kaplan, 2007; Xiao & Felleman, 2004; Xiao et al., 2003; Xiao et al., 1999). For instance, based on Felleman et al.'s (1997) work, Figure 3.4 illustrates that V4, along with the lower-level cortical areas V1 and V2, has separate neural compartments for shape and surface processing of visual objects. Regarding chromatic surface processing, Xiao et al. (2003) found that the thin stripes in V2, which receive input from V1's wavelength-selective blob areas, contain functional wavelength-specific subregions in which variations of stimulus wavelength are systematically mapped onto varying locations within the subregions. Subsequently Xiao et al. (2007) showed that cortical wavelength maps exist as early as in the V1 blob areas and suggested that they presumably provide input to the spatially more extensive V2 thin-stripe wavelength maps. Regarding achromatic surface processing, Wang et al. (2007) additionally found that neurons, located adjacent to the wavelength maps within the V2 thin stripes, respond

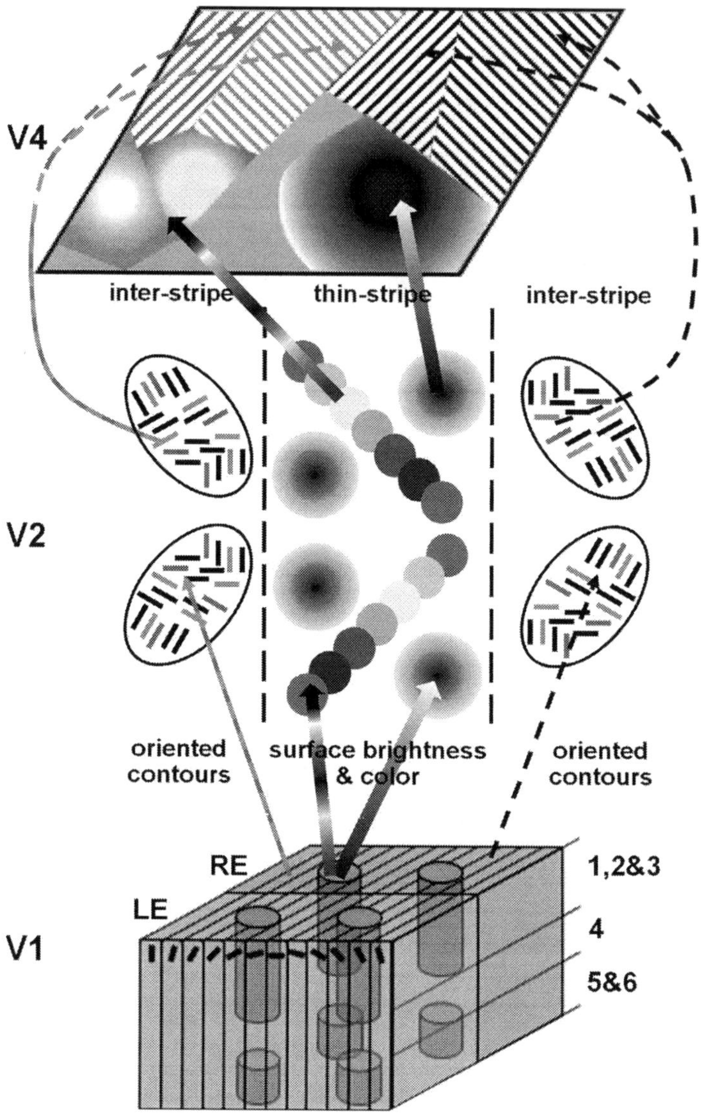

FIGURE 3.4 An elaboration of the ventral pathway.

Here, area V1 is composed of numerous orientation columns and two blob columns for the left eye (LE) and right eye (RE). From V1 orientation columns, chromatic (grey dashed arrows) and achromatic (black dashed arrows) orientations project to interstripes of V2 and from there to their respective subregions in V4; and from V1 blobs chromatic (narrow grey gradated arrows) and achromatic (wide grey gradated arrows) surface properties project respectively to the chromatic maps and achromatic regions of V2 and from there to their respective subregions in V4. [From Breitmeyer (2014), and from Wang, Xiao, & Felleman (2007) by permission of Oxford University Press.]

differentially to positive (light-on-dark) and negative (dark-on-light) luminance contrast. As suggested by Wang et al. (2007), the thin stripes of V2, adjacent to the interstripe orientation maps, thus comprise neural modules for processing surface properties of visual stimuli; whereas neurons found in V2 interstripes respond selectively to contour or edge orientation of stimuli and receive input from V1 inter-blob areas.

Evidence for Separate but Interactive Form-Processing and Surface-Processing Systems in Human Vision

Besides the neurophysiological and neuroanatomical evidence for neural substrates of the BCS and FCS systems in monkey visual cortex, several studies of human subjects also support the BCS-FCS distinction. Snow, Strother, Coros and Culham (2012) recently reported brain-imaging (fMRI) results showing independent color and form processing in the human ventral cortical pathway. Studies of neurological patients have shown that the shape and surface properties of visual objects are processed by dissociable cortical systems in humans. Several of these studies (Barbur, et al., 1994; Heywood et al., 1991; Kentridge et al., 2004) found that, despite loss of phenomenal hue perception of surfaces, achromatopsic (cortically color-blind) patients can detect contours defined by isoluminant chromatic contrasts. Moreover, consistent with Wang et al.'s (2007) findings in monkeys indicating the existence of separate luminance and color processing within V2 thin-stripe surface-processing modules, Heywood, Wilson and Cowey (1987) found that an achromatopsic patient who, while grossly impaired in discriminating isoluminant hues, was able to discriminate achromatic greys. Regarding the separate processing of surface and shape properties, Zeki, Aglioti, McKeefry and Berlucchi (1999) similarly showed that a patient, though all but form blind, was able to identify the colors of objects. The combined studies of neurological patients and of normal observers in Ganzfeld stimulation thus indicate that in human vision, like in that of primates, form- and surface-processing systems can be separated from each other.

To construct a veridical object representation in which surface properties fill in the area enclosed by contours, the BCS and FCS must be able to communicate or interact appropriately with each other. Disturbances in these systems or in their interaction should therefore lead to distorted perceptions of object properties. Such perceptual distortions are demonstrated by some neurological patients whose interactions between contour-forming and surface-filling processes appear to be disturbed. In his review of deficits of color perception in neurological patients, Critchley (1965) reports cases in which the color of an object irradiates outward beyond its boundaries, sometimes at great distances from the boundaries of the object, and in which the boundaries of the object are often reported as very fuzzy or blurred. In other cases the color of an object is perceived as not adhering to its surface, but instead as free-floating in space, in a plane distinct from that of the object and usually phenomenally located somewhere between the object and the patient. Using a variation of binocular-rivalry suppression known as interocular continuous flash suppression (Tsuchiya & Koch, 2005), Hong and Blake (2009) reported related phenomena in a study of normal observers. In this study rapidly changing achromatic (grey) Mondrian patterns were flashed to one eye while a stationary chromatic bar was presented to the other eye. Although observers failed to see the oriented shape of the bar, they did report its color in a free-floating, cloudlike constellation. Both cases, the neurological symptoms and the experimental phenomena indicate that the form-processing BCS either does not construct the necessary two- and three-dimensional spatial constraints for the filling-in process of the color-FCS system or that such constraints are not communicated to the FCS system. Consequently, not only in the frontoparallel plane but also in depth, the visual processing of the object lacks proper boundaries within which color completion can occur.

So far, the discussion of the BCS and FCS was framed implicitly in a bottom-up hierarchical manner of cortical processing. Well known, however, is that successively higher cortical areas receiving feedforward projections from lower ones have extensive feedback projections to the lower areas (Barone et al., 2000; Felleman & Van Essen, 1991; Maunsell & Van Essen, 1983). Top-down re-entrant activation thus plays a major role in several theoretical approaches to visual object perception (Ahissar & Hochstein, 2004; Bar et al., 2006; Lamme, 2010; Seth et al., 2004; Zeki, 1993). This indicates that visual object perception, and visual perception in general, depend on the dynamic neural interactions between bottom-up feedforward and top-down feedback cortical processing.

Neuronal Approaches to Interactions Between Bottom-Up and Top-down Visual Processing

As shown by Pomerantz and Cragin (2015), the spatiotemporal arrangement of perceptual components or parts can in a bottom-up manner dramatically influence the emergence in perception of Gestalt-like configurations. Conversely, as shown by Naomi Weisstein's extensive research, the Gestalt- or object-like configuration can in a top-down manner affect how its parts are processed and perceived. The above evidence reported in studies of monkey and human subjects thus allows us to relate the TOTE and analysis-by-synthesis concepts proposed in the 1960s – and the much earlier notion phrased by Helmholtz (1867) in terms of *unbewusste Schluss* (unconscious inference) and more recently in terms of perceptual postulate by Gregory (1997) – to plausible neural correlates. Research and modelling in recent decades has revealed plausible neural models of hierarchical bottom-up processing but also of top-down (reverse-hierarchical) influences. These rely on the distinction between the P and M pathways in the visual system. According to the frame-and-fill approach taken, for instance, by Bar et al. (2006), the global structure of an object (frame) is projected via the fast, dorsal cortical M pathway to the prefrontal cortex (PFC), where it generates in neural space a set of possible objects, each of which can be considered an analogue of a perceptual hypothesis. However, an alternate frame-and-fill approach, which takes into consideration 1) that M neural activity is also found in the ventral cortical pathway (Krubitzer & Kaas, 1990; Nealy & Maunsell, 1994; Sawatari & Callaway, 1996) and 2) that there are fast direct connections from posterior occipital to anterior temporal and frontal cortical regions (Catani et al., 2002, 2003), assumes that the rapid activation of global representations of objects in the PFC could also occur via the ventral route. In either approach, the information in the PFC's possible-objects space (i.e., the set of expected objects) could project down to the inferotemporal cortex. Via several iterative feedback-feedforward loops cascading through intermediate levels all the way to the striate V1 level, the best match between lower-level processing and several higher-level representations of possible objects determines which evidence-based "inference to the best object perception" wins out.

Bar et al.'s (2006) or similar neural frame-and-fill approaches would readily apply also to the filling-in effect giving rise to the perception of several illusory percepts, among them of course the illusory phantom gratings. Evidence for filling-in is found in such visual cortical areas as V1 and V2 (Meng et al., 2005). Although not ruling out the possible role of intra-areal neural connectivities in these early cortical sites (Meng et al., 2005), significant contributions to the overall filling in also could be made by inter-areal, top-down influences originating in higher-level extrastriate cortical areas (Angelucci et al., 2002; Mendola, 2003; Mumford, 1994).

The frame-and-fill approach is an instantiation of a more general and ubiquitous predictive coding, Bayesian inference scheme (for details see Cardin et al. (2011), Lee & Mumford (2003), Shipp (2016) and Summerfield & de Lange (2014)). The importance of perceptual expectations guided by context and the resulting predictive coding are important computational strategies not only deployed, and studied, in the vision laboratory but also deployed in everyday extralaboratory visual exploration of the environment. Two recent findings are highly relevant here. For one, Williams et al. (2008) report fMRI findings showing that human foveal retinotopic cortex contained information about objects presented in the periphery. They ruled out the hypothesis that automatic predictions of imminent saccades to peripheral stimuli might lead to predictive signals in foveal retinotopic cortex because, for one thing, the foveal information about peripheral targets built up slowly over a block of trials. However, the "objects" used by Williams et al. were 2-D renditions of 3-D nonsense figures. Such objects would not have a prior representation in a durable long-term visual store. As noted by Williams et al., it would take many trials to learn to discriminate and identify these objects and establish their durable visual memory. Only then can these memorial representations become part of the possible object-space activated by their coarse peripheral information. In everyday, extralaboratory settings, identity of highly familiar and contextually constrained objects in our visual environment would be more akin to Wiliams et al.'s nonsense objects having attained familiarity at the end of a block of trials than the unfamiliar counterparts at the beginning of the block. For that reason, Williams et al.'s (2008) rejection of the automatic predictive coding pertains only to their particular paradigm. In fact, it is quite reasonable to claim that once learning of the nonsense figures is attained their finding supports a hypothesis of automatized, in effect automatic, predictive coding based on a bottom-up activation of a predictive object space such as proposed by Bar et al. (2006). Second, dovetailing nicely with Williams et al.'s (2008) findings, Vetter, Edwards and Muckli (2012) showed that predictive signals can be transferred across saccades. Since saccades typically move the eyes to fixate a prior peripheral object, upon fixation, the object, now projecting its image onto the foveal area of the retina, would be processed in the foveal retinotopic cortex and where its detailed feedforward signals would be used to test the predictive perceptual hypotheses generated by the prior coarsely encoded peripheral object.

In my opinion, a frame-and-fill approach such as the one outlined above also provides an apt metaphor for Naomi Weisstein's lasting contribution to empirical and theoretical research on visual perception. Her work some 30+ years ago can be regarded, on the one hand by some, as contributing to the important experimental details that constitute the contents of current theoretical frameworks of visual object recognition yet, on the other hand by others including me, also as establishing a paradigmatic conceptual frame which since then has been filled in detail by accumulating findings and developments in visual cognition and visual neuroscience.

A Personal Note

I first met Naomi Weisstein in the fall of 1972 when I gave a colloquium in the Department of Psychology, SUNY Buffalo, where she had just moved from Loyola University Chicago. From then onward we hit it off fabulously as colleagues and friends. I next had the pleasure of running into her at Bell Labs where I was collaborating with Bela Julesz and Naomi was collaborating with Charlie Harris. I always considered her to be a very energetic, creative and inventive mind, laying the psychophysical and modelling groundwork for significant parts of what is being investigated by current vision and perception scientists as well visual neuroscientists. Although her affliction with chronic fatigue syndrome in the early 1980s left her bedridden until she passed away, she remained a vibrant and productive contributor to, and commentator on, vision science. Along with many others, I miss her spunky, enthusiastic and humorous spirit. I never had the occasion to meet her husband Jessie, who stayed by her side and advocated for her in many important ways and through critical transitions of her illness until she passed away. I admire his loyalty, patience and courage.

References

Ahissar, M., & Hochstein, S. (2004). The reverse hierarchy theory of visual perceptual learning. *Trends in Cognitie Sciences, 8*, 457–464.

Angelucci, A., Levitt, J. B., Walton, E. J. S., Hupé, J.-M., Bullier, J., & Lund, J. S. (2002). Circuits for local and global signal integration in primary visual cortex. *Journal of Neuroscience, 22*, 8633–8646.

Bar, M., Kassam, K. S., Ghuman, A. S., Boshyan, J., Schmid, A. M., Dale, A. M., Hämäläinen, M. S., Marinkovic, K., Schacter, D. L., Rosen, B. R., & Halgren, E. (2006). Top-down facilitation of visual recognition. *Proceedings of the National Academy of Science USA, 103*, 449–454.

Barbur, J. L., Harlow, A. J., & Plant, G. (1994). Insights into the different exploits of colour in the visual cortex. *Proceedings of the Royal Society B, 258*, 327–334.

Baron, J., & Thurston, I. (1973). An analysis of the word superiority effect. *Cognitive Psychology, 4*, 207–228.

Barone, P., Batardiere, A., Knoblauch, K., & Kennedy, H. (2000). Laminar distribution of neurons in extrastriate areas projecting to visual areas V1 and V4 correlates with the hierarchical rank and indicates the operation of a distance rule. *Journal of Neuroscience, 20*, 3263–3281.

Breitmeyer, B. G. (2014). *The Visual (Un)conscious and its (Dis)contents.* Oxford: Oxford University Press.

Brown, J. M., Gyoba, J., & May J. G. (2001). Stationary phantoms and grating induction with oblique inducing gratings: Implications for different mechanisms underlying the two phenomena. *Psychonomic Bulletin & Review, 8*, 278–283.

Cao, Y., & Grossberg, S. (2005). A laminar cortical model of stereopsis and 3D surface perception: Closure and da Vinci stereopsis. *Spatial Vision, 18*, 515–578.

Cardin, V., Friston, K. J., & Zeki, S. (2011). Top-down modulations in the visual form pathway revealed with dynamic causal modeling. *Cerebral Cortex, 21*, 550–562.

Catani, M., Howard, R. J., Pajevic, S., & Jones, D. K. (2002). Virtual in vivo interactive dissection of white matter fasciculi in the human brain. *NeuroImage, 17*, 77–94.

Catani, M., Jones, D. K, Donato, R., Ffytche, D. H. (2003). Occipito-temporal connections in the human brain. *Brain, 126*, 2093–2107.

Conway, B. R., Moeller, S., & Tsao, D. Y. (2007). Specialized color modules in macaque extrastriate cortex. *Neuron, 56*, 560–573.

Critchley, M. (1965). Acquired anomalies of colour perception of central origin. *Brain*, 88, 711–724.

Felleman, D. J., & Van Essen, D. C. (1991). Distributed hierarchical processing in the primate cerebral cortex. *Cerebral Cortex, 1*, 1–47.

Felleman, D. J., Xiao, Y., & McClendon, E. (1997). Modular organization of occipito-temporal pathways: Cortical connections between visual area 4 and visual area 2 and posterior inferotemporal ventral area in macaque monkeys. *Journal of Neuroscience, 17*, 3185–3200.

Fox, O. M., Harel, A., & Bennett, K. B. (2017). How configural is the configural superiority effect? A neuroimaging investigation of emergent features in visual cortex. *Frontiers in Psychology, 8*, 32.

Genter, C. R. II, & Weisstein, N. (1981). Flickering phantoms: A motion illusion without motion. *Vision Research, 21*, 963–966.

Gregory, R. (1997). Knowledge in perception and illusion. *Philosophical Transactions of the Royal Society of London B, 352*, 1121–1127.

Grossberg, S. (1994). 3-D vision and figure–ground separation by visual cortex. *Perception & Psychophysics, 55*, 48–120.

Grossberg, S. (2003). Filling-in the forms: Surface and boundary interactions in visual cortex. In L. Pessoa & P. De Weerd (eds) *Filling-in: From Perceptual Completion to Cortical Reorganization* (pp. 123–150). Oxford: Oxford University Press.

Grossberg, S., & Yazdanbakhsh, A. (2005). Laminar cortical dynamics of 3D surface perception: Stratification, transparency, and neon color spreading. *Vision Research, 45*, 1725–1743.

Helmholtz, H. v. (1867). *Handbuch der Physiologischen Optik*. Leipzig: L. Voss.

Herzog, M. H., & Clarke, A. M. (2014). Why vision is not both hierarchical *and* feed forward. *Frontiers in Computational Neuroscience, 8*, 135.

Herzog, M. H., Sayim, B., Chicherov, V., & Manassi, M. (2015). Crowding, grouping, and object recognition: A matter of appearance. *Journal of Vision, 15*(6):5, 1–18.

Heywood, C. A., Cowey, A., & Newcombe, F. (1991). Chromatic discrimination in a cortically colour blind observer. *European Journal of Neuroscience, 3*, 802–812.

Heywood, C. A., Wilson, B., & Cowey, A. (1987). A case study of cortical colour "blindness" with relatively intact achromatic discrimination. *Journal of Neurological and Neurosurgical Psychiatry, 50*, 22–29.

Hong, S. W., & Blake, R. (2009). Interocular suppression differentially affects achromatic and chromatic mechanisms. *Attention, Perception & Psychophysics, 71*, 403–411.

Kentridge, R. W., Heywood, C. A., & Cowey, A. (2004). Chromatic edges, surfaces, and constancies in cerebral achromatopsia. *Neuropsychologia, 42*, 821–830.

Kinoshita, M., & Komatsu, H. (2001). Neural representation of luminance and brightness of a uniform surface in the macaque primary visual cortex. *Journal of Neurophysiology, 86*, 2559–2570.

Komatsu, H. (2006). The neural mechanisms of perceptual filling-in. *Nature Reviews Neuroscience*, 7, 220–231.

Krubitzer, L., & Kaas, J. (1990). Convergence of processing channels in the extrastriate cortex of monkeys. *Visual Neuroscience, 5*, 609–613.

Lamme, V. A. F. (2010). How neuroscience will change our view of consciousness. *Cognitive Neuroscience, 1*, 204–240.

Lance, M., Weisstein, N., & Harris, J. R. (1982). Perceived depth vs. structural relevance in the object-superiority effect. *Perception & Psychophysics, 31*, 376–382.

Lee, T. S., & Mumford, D. (2003). Hierarchical Bayesian inference in the visual cortex. *Journal of the Optical Society of America A, 20*, 1434–1448.

Lu, H. D., & Roe, A. W. (2008). Functional organization of color domains in V1 and V2 of macaque monkey revealed by optical imaging. *Cerebral Cortex, 18*, 516–533.

McClelland, J. L., & Miller, J. (1979). Structural factors in figure perception. *Perception & Psychophysics, 26*, 221–229.

Maguire, W. M, & Brown J. M. (1987). The current state of research into visual phantoms. In S. Petry & G. E. Meyer (eds) *The Perception of Illusory Contours* (pp. 213–219). New York: Springer Verlag.

Maunsell, J. H. R., & Van Essen, D. C. (1983). The connections of the middle temporal visual area (MT) and their relationship to a cortical hierarchy in the macaque monkey. *Journal of Neuroscience, 3*, 2563–2586.

Mendola, J. (2003). Contextual shape processing in human visual cortex: Beginning to fill-in the blanks. In L. Pessoa & P. De Weerd (eds) *Filling-In: From Perceptual Completion to Cortical Reorganization* (pp. 38–58). Oxford: Oxford University Press.

Meng, M., Remus, D. A., & Tong, F. (2005). Filling-in of visual phantoms in the human brain. *Nature Neuroscience, 8*, 1248–1254.

Miller, G. A., Galanter, E., & Pribram, K. H. (1960). *Plans and the Structure of Behavior.* New York: Henry Holt & Company.

Mulvanny, P., Macarthur, R., Sekuler, R. (1982). Thresholds for seeing visual phantoms and moving gratings. *Perception, 11,* 35–46.

Mumford, D. (1994). Neuronal architectures for pattern-theoretic problems. In C. Koch & J. Davis (eds) *Large-Scale Neuronal Theories of the Brain* (pp. 125–152). Cambridge, MA: MIT Press.

Nealy, T.A., & Maunsell, J.H.R. (1994). Magnocellular and parvocellular contributions to the responses of neurons in macaque striate cortex. *Journal of Neuroscience, 14,* 2069–2079.

Neisser, U. (1967). *Cognitive Psychology.* New York: Appleton-Century-Crofts.

Pomerantz, J. R., & Cragin, A. I. (2015). Emergent features and feature combination. In J. Wagemans (ed.) *The Oxford Handbook of Perceptual Organization* (pp. 88–107). Oxford: Oxford University Press.

Pomerantz, J. R., & Portillo, M. C. (2011). Grouping and emergent features in vision: Toward a theory of basic Gestalts. *Journal of Experimental Psychology: Human Perception and Performance, 37,* 1331–1349.

Reicher, G.M. (1969). Perceptual recognition as a function of meaningfulness of stimulus material. *Journal of Experimental Psychology, 81,* 275–280.

Roe, A. W., & Ts'o, D. Y. (1999). Specificity of color connectivity between primate V1 and V2. *Journal of Neurophysiology, 82,* 2719–2730.

Rosenbach, O. (1902) Zur Lehre von den Urtheilstäuschungen. *Zeitschrift für Psychologie, 29,* 434–448.

Sawatari, A., & Callaway, E. M. (1996). Convergence of magno- and parvocellular pathways in layer 4B of macaque primary visual cortex. *Nature, 380,* 442–446.

Seth, A. K., McKinstry, J. L., Edelman, G. M., & Krichmar, J. L. (2004). Visual binding through reentrant connectivity and dynamic synchronization in a brain-based device. *Cerebral Cortex, 14,* 1185–1199.

Shannon, C. E., & Weaver, W. (1949). *The Mathematical Theory of Communication.* Urbana, IL: University of Illinois Press.

Shipp, S. (2016). Neural elements for predictive coding. *Frontiers in Psychology, 7,* 1792.

Sincich, L. C., & Horton, J. C. (2005). The circuitry of V1 and V2: Integration of color, form, and motion. *Annual Review of Neuroscience, 28,* 303–326.

Snow, J. C., Strother, L., Coros, A. M., & Culham, J. C. (2012). How independent are form and color in the ventral visual pathway? *Journal of Vision 12*(9), 510.

Summerfield, C. & de Lange, F. P. (2014). Expectation in perceptual decision making: neural and computational mechanisms. *Nature Reviews Neuroscience, 15,* 745–756.

Tsuchiya, N., & Koch, C. (2005). Continuous flash suppression reduces negative afterimages. *Nature Neuroscience, 8,* 1096–1101.

Turing, A. M. (1950). Computing machinery and intelligence. *Mind, 49,* 433–460.

Tynan, P., & Sekuler, R. (1975). Moving visual phantoms: A new contour completion effect. *Science, 188,* 951–952.

Vetter, P., Edwards, G., & Muckli, L. (2012). Transfer of predictive signals across saccades. *Frontiers in Psychology, 3,* 176.

von Neumann, J. (1958). *The Computer and the Brain.* New Haven/London: Yale University Press.

Wang, Y., Xiao, Y., & Felleman, D. J. (2007). V2 thin stripes contain spatially organized representations of achromatic luminance change. *Cerebral Cortex, 17,* 116–129.

Weisstein, N. (1968). A Rashevsky-Landahl neural net: Simulation of metacontrast. *Psychological Review, 75,* 494–521.

Weisstein, N., & Harris, C. S. (1974). Visual detection of line segments: An object-superiority effect. *Science, 186*, 752–755.

Weisstein, N., Maguire, W., & Berbaum, K. (1977). A phantom-motion aftereffect. *Science, 198*, 955–958.

Weisstein, N., Maguire, W. & Williams, M. C. (1982a). The effects of perceived depth on phantoms and the phantom motion aftereffect. In J. Beck (ed.) *Organization and Representation in Perception* (pp. 235–249). Hillsdale, NJ: Lawrence Erlbaum.

Weisstein, N., Ozog, G., & Szoc, R. (1975). A comparison and elaboration of two models of metacontrast. *Psychological Review, 82*, 325–343.

Weisstein, N., Williams, M. C., & Harris, C. S. (1982b). Depth, connectedness, and structural relevance in the object-superiority effect: Line segments are harder to see in flatter patterns. *Perception, 11*, 5–17.

Wheeler, D. D. (1970). Processes in word recognition. *Cognitive Psychology, 1*, 59–85.

Wiener, N. (1948). *Cybernetics, or Control and Communication in the Animal and the Machine*. Cambridge, MA: MIT Press.

Williams, M. A., Baker, C. I., Op de Beeck, H. P., Shim, W. M., Dang, S., Triantafyllou, C., & Kanwisher, N. (2008). Feedback of visual object information to foveal retinotopic cortex. *Nature Neuroscience, 11*, 1439–1445.

Wong, E., & Weisstein, N. (1982). A new perceptual context-superiority effect: Line segments are more visible against a figure than against a ground. *Science, 218*, 587–589.

Wong, E., & Weisstein, N. (1987). The effects of flicker on the perception of figure and ground. *Perception & Psychophysics, 41*, 440–448.

Xiao, Y., & Felleman, D. J. (2004). Projections from primary visual cortex to cytochrome oxidase thin stripes and interstripes of macaque visual area 2. *Proceedings of the National Academy of Science USA, 101*, 792–804.

Xiao, Y., Wang, Y., & Felleman, D.J. (2003). A spatially organized representation of colour in macaque cortical area V2. *Nature, 421*, 535–539.

Xiao, Y., Zych, A., & Felleman, D. J. (1999). Segregation and convergence of functionally defined V2 thin stripe and interstripe compartments projecting to area V4 of macaques. *Cerebral Cortex, 9*, 792–804.

Xiao, Y., Casti, A., Xiao, J., & Kaplan, E. (2007). Hue maps in primate striate cortex. *NeuroImage, 35*(2), 771–786.

Zeki, S. (1993). *A Vision of the Brain*. Oxford: Blackwell.

Zeki, S., Aglioti, S., McKeefry, D., & Berlucchi, G. (1999). The neurological basis of conscious color perception in a blind patient. *Proceedings of the National Academy of Science USA, 96*, 14124–14129.

4

VISUAL PHANTOM ILLUSION AS AN INTEGRATIVE PRODUCT OF EARLY VISUAL PROCESSING AND HIGHER-ORDER PERCEPTUAL ORGANIZATION

Jiro Gyoba[1], Kenzo Sakurai[2] and Akiyoshi Kitaoka[3]
[1]Tohoku University, Japan, [2]Tohoku Gakuin University, Japan, [3]Ritsumeikan University, Japan

Introduction

Naomi Weisstein has vigorously applied psychophysical measurements, such as differential adaptation or masking, to the investigation of higher-order perceptual processes including surface completion (Weisstein, 1970, 1973), three-dimensional object perception (Weisstein & Harris,1974), and figure-ground organization (Klymenko & Weisstein, 1986; Weisstein & Wong, 1986).

One of the most pioneering hypotheses underlying her studies is 'neural symbolic activity.' This postulates that the same units involved in early visual processing are also engaged in higher-order perceptual processes (1970, 1973). For example, Weisstein, Montalvo, and Ozog (1972) showed participants a grating that was partially blocked from view by a perspective drawing of a cube. They found that adaptation occurred not only to the visible portions of the grating but also to the invisible portions blocked from view. Based on the neural symbolic activity hypothesis, this finding can be interpreted as follows: early visual system's activity (such as extracting spatial frequency) can be modified by the higher-order processing of the overall configuration, corresponding to our perceptual interpretation that the blocked grating appears to continue behind the cube as a background. Gyoba (1979) confirmed that the adaptation effect to invisible gratings occurred when the occluding figure was placed stereoscopically in front of the surrounding grating, and did not occur when the occlude was stereoscopically behind the surround.

The existence of the neural activity such as that predicted by Weisstein is demonstrated in a neurophysiological study by Sugita (1999) who found that the V1 neuronal firings responding to a bar occurred even when the mid-part of the bar was occluded by a stereoscopically-raised square, and did not occur with the stereoscopically-downed squares.

In figure-ground perception, a continuous background is completed behind the figure, although it is not visible. Kanizsa (1979) called this general perceptual function as 'amodal completion'. On the other hand, there is another type of phenomenon called 'modal completion' in which the filled-in parts are actually visible at locations where there are no physical features in the image. One of the typical examples of modal completion phenomena is the visual phantom illusion. When the central part of a vertical grating is blocked by a horizontal black occluder, the top and the bottom sections of the grating visually appear to continue in front of the occluder.

A modal completion phenomenon of this kind was initially reported by Rosenbach (1902). Tynan and Sekuler (1975) demonstrated it clearly using moving patterns on a CRT monitor, naming the illusion 'visual phantoms.' Since then, a series of studies has been carried out and many of the psychophysical characteristics of the phantom phenomenon have been revealed. Phantom visibility is affected by several variables. For example, when the spatial frequency of inducing gratings is high and the width of the occluder is thick, phantoms are not visible (Tynan & Sekuler, 1975; Genter & Weisstein, 1981; Gyoba, 1983). Phantoms are also enhanced by the motion or the flicker of the inducing gratings (Tynan & Sekuler, 1975; Genter & Weisstein, 1981), yet the illusion can be perceived with stationary gratings (Gyoba, 1983). The optimal occluder luminance for seeing phantoms is close to either the minimum or the maximum luminance of the inducing gratings (Sakurai & Gyoba, 1985). Finally, when the occluder is seen in front of the inducing pattern by binocular disparity or figural depth cues, phantom visibility is reduced (Weisstein et al., 1982; Brown & Weisstein, 1991).

One of the intriguing problems of visual phantoms is their relation to the grating induction effect demonstrated by McCourt (1982). He found that a vertical inducing grating surrounding a horizontal test field of average luminance produced the appearance of an opposite phase grating within the homogeneous test field. Since then, a long controversy has existed between McCourt and the phantom researchers. McCourt (1994) insisted that the visual phantom could be explained by the grating induction effect based on his finding that the visible induced grating on the occluded area actually consists of a grating in opposite phase to the inducing grating. However, the conditions for McCourt's out-of-phase grating induction are quite different from those for the phantom gratings. The optimal luminance of occluder (test field in McCourt's terminology) for seeing the grating induction effect is at the average luminance of grating, while at the minimum or maximum luminance in the case of the phantom gratings (Sakurai & Gyoba, 1985). Moreover, the visibility of phantoms is higher under low luminance inducing grating (Gyoba, 1994a) and the phantom has a figural fidelity to the oblique inducing grating (Gyoba, 1994b; Brown et al., 2001), while the opposite tendencies are observed in the case of the grating induction effect (McCourt, 1990; Brown et al., 2001).

In this chapter, we introduce a new illusion, phantom induced blindness (PIB) in which high intensity yellow dots placed on the stationary inducing gratings perceptually disappear when the phantom gratings are perceived. Then, we show that both moving and stationary visual phantoms can be induced by second-order components (texture, color, contrast-modulation, etc.) and robust against the luminance change of the occluder. Finally, we proposed a two-stage model and discussed visual phantoms from the viewpoint of Weisstein's neural symbolic activity hypothesis. This describes the phenomenon of surface completion as a modification of early visual processing activities based on perceptual processing of overall configuration.

A New Illusion That Differentiates the Phantoms From the Grating Induction: Phantom Induced Blindness

Bonneh, Cooperman, and Sagi (2001) reported that salient visual targets alternately disappear from awareness and reappear when they are superimposed on a moving distractor pattern. They called this phenomenon 'motion-induced blindness (MIB)' and discussed the underlying mechanism from the view point of attentional competition between the targets and the distractor. Later, other explanations were proposed emphasizing the importance of surface completion or filling-in processes, by indicating that MIB increased when the depth ordering between target and distractor patterns was consistent with the perceptual occlusion (Graf et al., 2002; Hsu et al., 2006).

Recently we have found that a similar disappearance phenomenon occurs in the visual phantom illusion. When high intensity yellow dots were placed on the stationary inducing gratings (Figure 4.1), one or sometimes all of the dots would often perceptually disappear when the black phantom grating connecting the top and the bottom sections of the grating appeared in front of the occluder. We named this effect 'phantom-induced blindness (PIB).' One might suspect that PIB may be due to the Troxler effect, in which small targets located at the periphery of the visual field tend to disappear. We have conducted preliminary experiments in order to indicate that the Troxler effect is not the main cause for PIB, by comparing the disappearance duration under the condition suitable for producing the phantoms (Figure 4.1) with the duration under the condition suitable for the grating induction effect (Figure 4.2), as described below.

The stimuli shown in Figures 4.1 and 4.2 were created and presented on a CRT monitor. The rectangular stimulus region subtended 18.5 (H) × 27.5 (W) degree in visual angle. The 0.21 cpd sine-wave inducing grating with a contrast of 0.90 and a mean luminance of 13.5 cd/m^2 was presented in the rectangular stimulus region. The height of the occluder was 4.5 degrees and its luminance was 0.50 cd/m^2 in the case of the phantom condition and 13.5 cd/m^2 in the grating

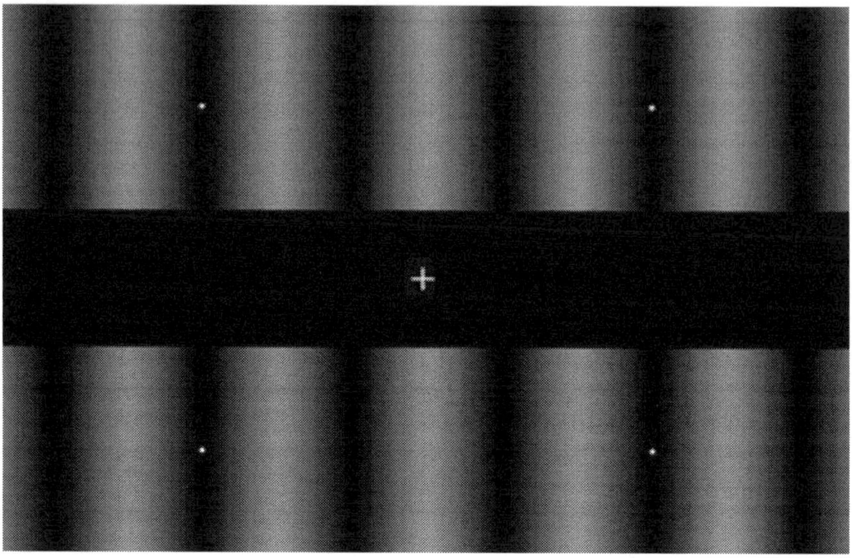

FIGURE 4.1 A demonstration of Phantom Induced Blindness (PIB).

If you stare at the cross-like figure located at the center of the black occluder, when the black phantom gratings are perceived in the front of the occluder, one or sometimes all of the four dots will often disappear.

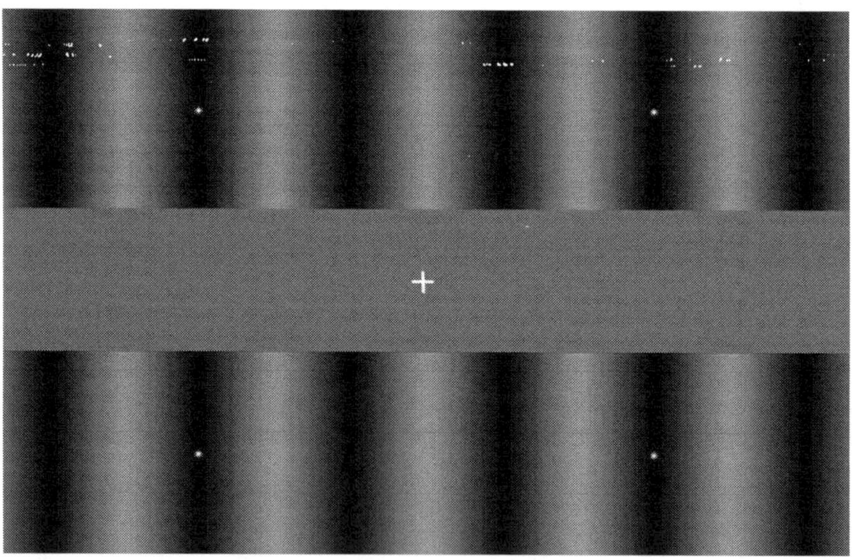

FIGURE 4.2

When the luminance of the occluder is near to the average luminance of the inducing gratings, the grating induction effect (GI) becomes dominant, resulting in far less frequent disappearance of the dots.

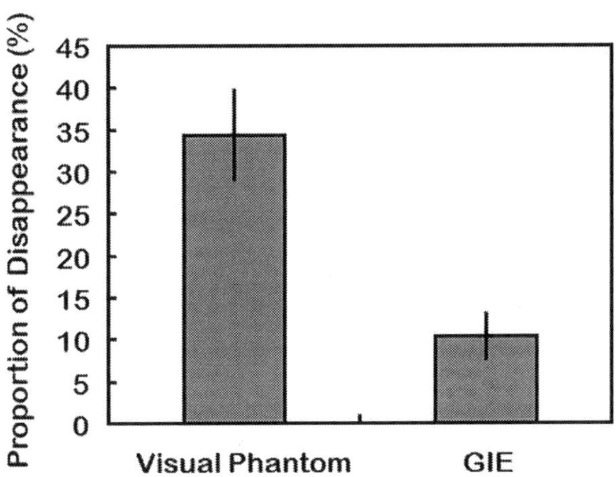

FIGURE 4.3 Proportion of cumulative disappearance duration in the phantom and the grating induction (GIE) conditions.

Error bars denote standard errors.

induction condition. At the center of the occluder, a white cross-like figure was presented as a fixation point. Four high-luminance yellow dots (63 cd/m^2, 0.3 degrees in diameter) were placed at 9 degrees away from the fixation figure. Three viewers with corrected-normal vision participated in the observation in a dark room with a chinrest. Viewing was performed through a single eye. They were asked to press a key while either of the four dots disappeared and to release the key while all dots reappeared in a 30 sec observation trial. Six observation trials were conducted for each of the phantom and the grating induction conditions.

The result (Figure 4.3) showed that at least one of the four dots disappeared during about 35 percent of the total viewing time in the case of visual phantom, while disappearance occurs only 10 percent of the time in the grating induction condition. This is despite the fact that the four dots were separated with the same distance from the fixation in both conditions. This fact indicated that the stimulus eccentricity (the important factor of the Troxler effect) is not the main cause for the disappearance.

We have conducted another preliminary experiment using eight observers by manipulating the special frequency of the inducing grating. It is well known that the phantom visibility decreases with the increase of special frequency (Tynan & Sekuler, 1975; Genter & Weisstein, 1981; Gyoba, 1983). In the phantom stimulation like Figure 4.1, we found that the proportion of the disappearance duration monotonically decreased from approximately 35 to 15 percent with the special frequency increase from 0.21 to 1.0 cpd. These results suggest that the disappearance phenomenon could be related to the phantom visibility, while we further need to investigate the effects of other parameters such as the contrast, average luminance, or orientation of gratings.

When the phantoms are perceived, the black bars of the grating appear to continue in the front of the black occluder. The surface completed by visual phantoms could hide the dots and produce perceptual blindness. Such a shift of surface depth ordering by perceptual completion may be consistent with recent explanations for MIB (Graf, Adams, & Lages 2002; Hsu, Yeh, & Kramer, 2006), suggesting that a common mechanism might be involved in the MIB and PIB.

Visual Phantoms Induced by Non-Luminance-Based Components

In the previous studies of visual phantoms and grating induction, luminance-based stimulation has been used to investigate the mechanisms behind the effects. We found, however, visual phantoms are induced by second-order components, that is, non-luminance-based stimulus components (Kitaoka et al., 2006). The following are three examples of second-order static phantoms.

Figure 4.4 shows the first example of inducing second-order components and the induced phantoms (contrast-modulated-dependent phantoms, or envelope phantoms). The contrast modulated random-dot pattern is used as the inducing gratings, and the occluder has the mean luminance of the inducing pattern. Even though there is not any difference among locally averaged luminance values in the whole pattern, the lowest contrast regions of the inducing contrast-modulation are perceived to be continuous on the occluder.

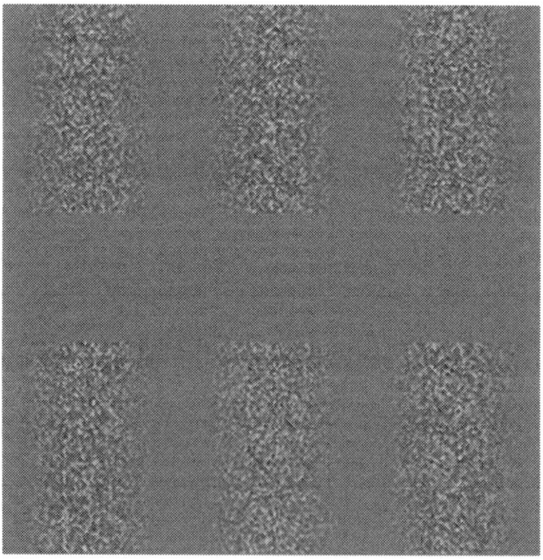

FIGURE 4.4 Phantoms induced by contrast-modulated random-dot patterns.

The intriguing point is that the luminance level of the occluder matches the mean luminance of the random-dot patterns. Visual phantoms induced by luminance-modulated gratings are most vivid when the luminance of the occluder matches the maximum/minimum luminance of the inducing gratings (Sakurai & Gyoba, 1985). On the contrary, in the second-order phantoms, the inducing pattern contains only a maximum/minimum contrast profile instead of a maximum/minimum luminance profile. Since the contrast of the occluder with flat luminance is 0, the minimum contrast (0) region in the contrast-modulated pattern appears to be continuous over the occluder which has the same luminance as the mean luminance of the inducing random-dot patterns. Considering the fact that the regions with equal luminance are perceived to be continuous in both luminance-modulated and contrast-modulated phantoms, 'assimilation in borderless fields' demonstrated by K. Koffka (Spillmann et al., 2006) could account for both types of phantoms.

The second example of second-order phantoms is phantom contours produced by a Craik-O'Brien-Cornsweet illusion pattern, as shown in Figure 4.5. The inducing pattern in this figure has only a local narrow luminance profile which produces contours. These contours are perceptually continuous on the occluder, though most of the areas have equal luminance (Kitaoka et al., 2006).

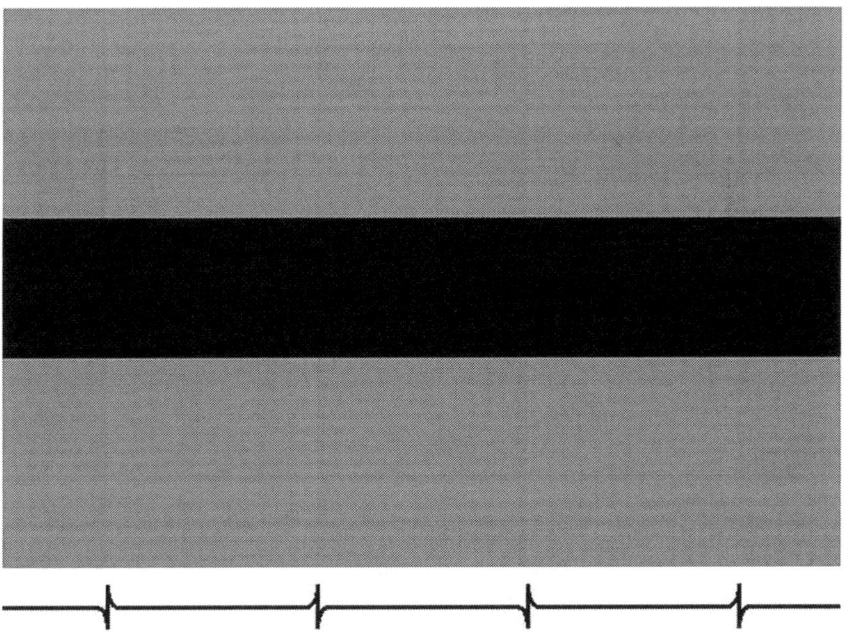

FIGURE 4.5 Phantom contours induced by Craik-O'Brien Cornsweet pattern.

The line at the bottom represents the luminance profile.

We could say that this phantom is induced by a difference of 'lightness' that is also a product of higher-order visual processing.

The third example of second-order phantoms is shown in Figure 4.6. The inducing pattern consists of a black horizontal occluder and plaids oriented 45 degrees which are contrast-modulated by square wave (envelope). Illusory contours of inducing patterns are perceived to be continuous on the horizontal black occluder.

The visibilities of these three examples of phantoms are enhanced if the inducing patterns were moved. A moving version of the first example is shown as Movie 1. Intriguingly, in a moving version of the third example shown as Movie 2, not only illusory contours but also dim illusory plaids are perceived on the occluder. Note that the plaid pattern (carrier) is not moving per se in the moving stimulus configuration.

FIGURE 4.6 Phantom contours and dim illusory plaids on the occluder induced by contrast-modulated plaids.

This type of second-order phantoms might be a variant of phantoms described by Weisstein and her colleagues (Weisstein et al., 1977; Weisstein & Harris, 1980). They created an inducing stimulus by moving vertical black stripes on a complex texture (arrays of Xs), and placed a black horizontal occluder. They found that observers perceived phantom contours and dim columns of Xs crossing the black horizontal occluder (see Patrick Cavanagh's chapter). Their complex stimulus configuration which created illusory moving contours is quite close to the square-wave contrast-modulation in the third example of our demonstration. The dimly perceived columns of Xs are likely to be a product of filling-in process because it is difficult to account for the completion of complex patterns solely by low-level luminance-based mechanisms.

Movies

To view movies for Chapter 4 go to https://research.franklin.uga.edu/visual-perception-laboratory/chapter-4

Movie 1 Moving version of phantoms induced by contrast-modulated random-dot patterns. The optimal condition is to view this movie in a dark room with the stimulus size subtending around 8° of visual angle. Make sure it loops.

Movie 2 Moving version of phantom contours and dim illusory plaids on the occluder induced by contrast-modulated plaids. The optimal condition is to view this movie in a dark room with the stimulus size subtending around 8° of visual angle. Make sure it loops.

Two-stage model

It was originally believed that the brightness induction in visual phantoms is in phase with the inducing grating (Mulvanny et al., 1982; Weisstein et al., 1982), as graphically exaggerated in Figure 4.7a, because the induced phantom grating appears to be continual with the inducing grating. This belief was questioned by McCourt (1994) who pointed out that the induced brightness on the occluder is in counterphase with the inducing grating, as graphically exaggerated in Figure 7b. McCourt's finding was confirmed in subsequent research (May et al., 1999).

McCourt (1994) suggested that this counterphase induction is a kind of grating induction (McCourt, 1982), which is one of the strongest brightness illusions. He proposed an explanation of dark phantoms in which the dark phase of the inducing grating is combined with the light phase of induced brightness on the black occluder because of brightness similarity between them. He then thought that counterphase brightness induction supports this perceptual continuation.

Grating induction, however, is quite different from visual phantoms in appearance since the standard grating induction pattern never gives perceptual continuation of the gratings in front of the occluder. To understand what makes this difference, we focused on perceptual transparency (Kitaoka et al., 2001). In this

FIGURE 4.7

(a) Exaggeration of 'in-phase' visual phantoms. In this figure, luminance modulation is given to the occluder in phase with the flanking gratings. (b) Exaggeration of 'counterphase' visual phantoms. In this figure, luminance modulation is presented on the occluder out of phase with the flanking gratings. In both cases, three dark columns appear to be continuous in front of the occluder.

phenomenon, adjacent areas that are similar in luminance tend to appear in front (Fuchs, 1923; Oyama & Nakahara, 1960, Metelli, 1974). We attributed the difference between the phantom and the grating induction effects to the classification of transparency (Anderson, 1997). Our idea is that visual phantoms reflect 'unique' transparency whereas McCourt's grating induction shows 'invalid' transparency.

Anderson (1997) phenomenologically classified perceptual transparency into two types: unique transparency and bistable transparency. In the former, a transparent surface is always perceived in front of the other surface, while in the latter, the perceived depth of two surfaces alternates and the surface in front appears to be transparent. These two types of perceptual transparency depend on the type of X-junctions. Unique transparency appears when contrast polarity along one edge is reversed over the X-junction and contrast polarity along the other edge is preserved over the X-junction (Figure 8a). On the other hand, bistable transparency appears when contrast polarity along both edges is preserved over the X-junction (Figure 8b). If contrast polarity along both edges is reversed over the X-junction, invalid transparency appears (Figure 8c). According to Anderson (1997), this idea was first proposed by Adelson and Anandan (1990). For further research on this issue, see Kitaoka (2005).

It is evident that the configuration of grating induction corresponds to that of invalid transparency. This is the main reason why the grating induction never shows perceptual continuation of the grating, since those edges are not grouped by transparency. On the other hand, the configuration of visual phantoms renders a series of configurations of unique transparency. In unique transparency, adjacent regions of the lower contrast always appear to be continual in front of the background when they are aligned with edges of different contrast polarities. This depth order is irreversible.

Figure 4.9 summarizes the two-stage model (Kitaoka et al., 2001, 2006) described so far. The lower stage reflects whether brightness induction is in phase or counterphase with inducing grating, whereas the higher stage displays the three types of transparency. When counterphase brightness induction is produced in the lower stage, the configural condition in the higher stage matches either unique transparency or invalid transparency. If it matches the former, visual phantoms (scotopic phantoms) are observed. If it follows the latter, grating induction is seen. When in-phase brightness induction is produced in the lower stage, the configural condition matches bistable transparency in the higher stage. In this case, photopic phantoms (Kitaoka et al., 1999) are observed.

Conclusion

Since the 1980s, there has been a long controversy between researchers working on phantom gratings and McCourt, who discovered the grating induction effect. As mentioned so far, we have introduced surface completion or a filling-in process revealed by the phantom-induced blindness (PIB) and the second-order

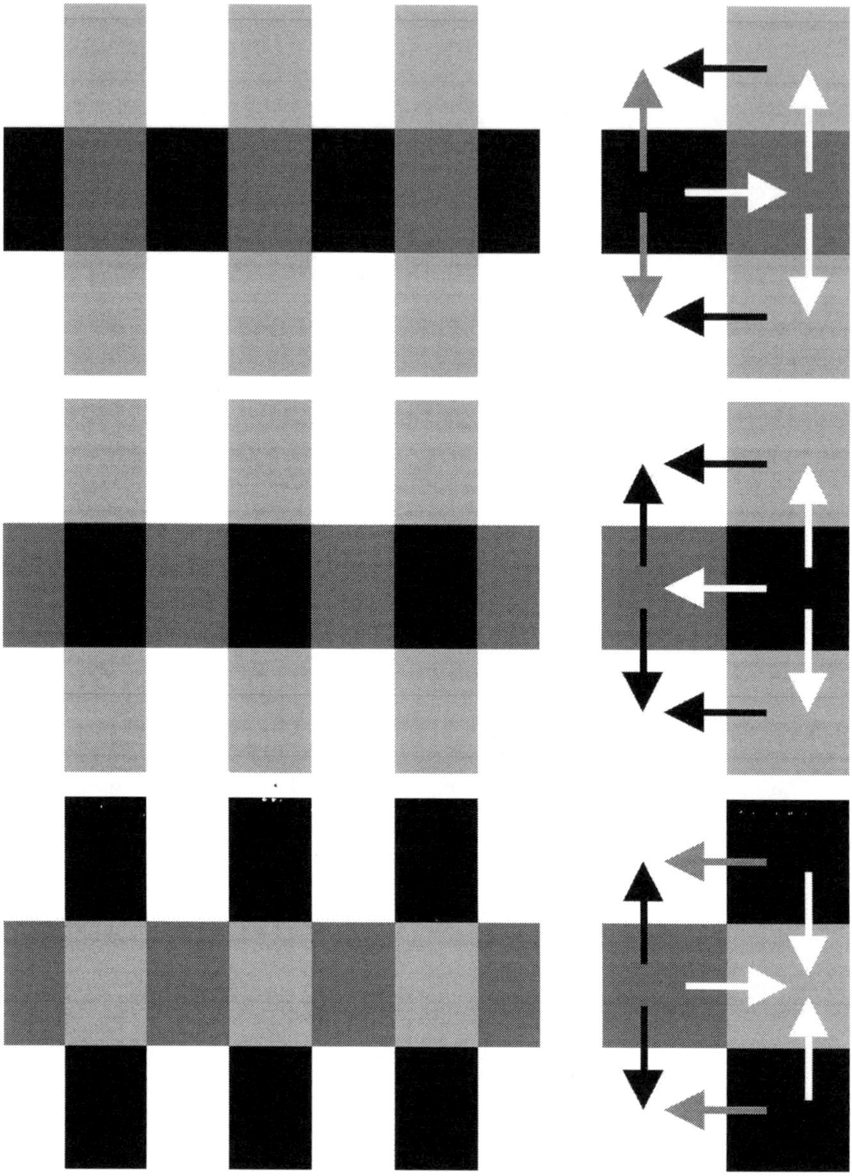

FIGURE 4.8 The phenomenological classification of perceptual transparency depending on contrast polarity along edges over X-junctions.

(a) Unique transparency, in which vertical gray rectangles appear to be transparent or translucent in front of the horizontal black rectangle. (b) Bistable transparency, in which vertical gray rectangles appear to be transparent in front of the horizontal gray rectangle, or the latter appears to be transparent in front of the former. (c) Invalid transparency, in which transparency is not perceived. Arrows indicate which region is brighter between two adjacent regions. [Figure reprinted from Kitaoka, Gyoba, & Sakurai (2006), with permission from Elsevier.]

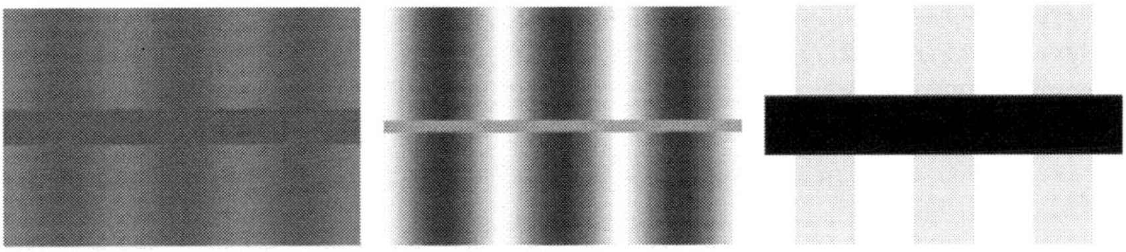

	Scotopic phantoms	Grating induction	Photopic phantoms
Lower stage (brightness induction)	counterphase	counterphase	in-phase
Higher stage (transparency)	unique	invalid	bistable

FIGURE 4.9 The two-stage model.

The lower stage reflects brightness induction while the higher one represents transparency. In the upper row, visual phantoms (scotopic phantoms) (left), grating induction (middle), and photopic phantoms (right) are depicted.

phantoms, and we proposed the two-stage model based on perceptual transparency. All of these findings provide evidence differentiating the visual phantom phenomenon from the grating induction effect and indicate the visual phantom illusion can be explained in terms of surface completion based on perceptual organization processes (Gyoba et al., 2016).

In line with this, Meng, Remus and Tong (2005) reported a fMRI evidence showing that visual phantoms lead to enhanced activity in early visual areas (V1 and V2) corresponding to the blank gap (occluder) region. They discussed that visual representations are recreated in 'zones of absence' and that the earliest stages of cortical processing (including the primary visual cortex) realize these representations of subjective content. Furthermore, Meng, Ferneyhough, and Tong (2007) proposed a two-stage model that consists of the fast processing corresponding to the blank region and the slow processing that produces suprathreshold levels of activity corresponding to the perceptual content of phantoms, in order to explain the time-consuming process of phantom perception.

While we must further consider whether there may be some common points or relationships between Meng's two-stage model and ours, most of the findings and models concerning visual phantom phenomenon could be interpreted within the framework proposed by Weisstein (1970). The activities of early visual processing could be modified dependent on higher-order perceptual organization, allowing the same units in early visual processing to engage also in perceptual interpretation.

References

Adelson, E. H., & Anandan, P. (1990) Ordinal characteristics of transparency. *Paper presented at the AAAI-90 Worksop on Qualitative Vision*, Boston, MA.

Anderson, B. L. (1997) A theory of illusory lightness and transparency in monocular and binocular images: The role of contour junctions. *Perception, 26*, 419–453.

Bonneh, Y. S., Cooperman, A., & Sagi, D. (2001) Motion-induced blindness in normal observers. *Nature, 411*, 798–801.

Brown, J. M., & Weisstein, N. (1991) Conflicting figure-ground and depth information reduces moving phantom visibility. *Perception, 20*, 155–165.

Brown, J. M., Gyoba J., & May, J. G. (2001) Stationary phantoms and grating induction with oblique inducing gratings: Implications for different mechanisms underlying the two phenomena. *Psychonomic Bulletin & Review, 8*, 278–283.

Fuchs, W. (1923) Untersuchungen über das simultane Hintereinandersehen auf derselben Sehrichtung. *Zeitshcrift für Psychologie, 91*, 145–235.

Genter, C. R. II., & Weisstein. N. (1981) Flickering phantoms: A motion illusion without motion. *Vision Research, 21*, 963–966.

Graf, E. W., Adams, W. J., & Lages, M. (2002) Modulating motion-induced blindness with depth ordering and surface completion. *Vision Research, 42*, 2731–2735.

Gyoba, J. (1979) Differential adaptation to invisible gratings blocked by stereoscopically raised planes. *Tohoku Psychologica Folia, 38*, 29–35.

Gyoba, J. (1983) Stationary phantoms: A completion effect without motion and flicker. *Vision Research, 23*, 205–211.

Gyoba. J. (1994a) Disappearance of stationary visual phantoms under high luminant or equiluminant inducing gratings. *Vision Research, 34*, 1001–1005.

Gyoba, J. (1994b) The visual phantom illusion under oblique inducing gratings. *Japanese Psychological Research, 36*, 182–187.

Gyoba, J., Sakurai, K., & Kitaoka, A. (2016) The visual phantom illusion revisited. *International Journal of Psychology, 51*, 1014.

Hsu, L. C., Yeh, S. L., & Kramer, P. (2006) A common mechanism for perceptual filling-in and motion-induced blindness. *Vision Research, 46*, 1973–1981.

Kanizsa, G. (1979) *Organization in Vision*. New York: Praeger.

Kitaoka, A. (2005) A new explanation of perceptual transparency connecting the X-junction contrast-polarity model with the luminance-based arithmetic model. *Japanese Psychological Research, 47*, 175–187.

Kitaoka, A., Gyoba, J., & Kawabata, H. (1999) Photopic visual phantom illusion: Its common and unique characteristics as a completion effect. *Perception, 28*, 825–834.

Kitaoka, A., Gyoba, J., & Sakurai, K. (2006) The visual phantom illusion: A perceptual product of surface completion depending on brightness and contrast. In: S. Martinez-Conde, S. L. Macknik, L. M. Martinez, J.-M. Alonso & P.U. Tse (Eds) *Visual Perception Fundamentals of Vision: Low and Mid-Level Processes in Perception, Progress in Brain Research*, Vol. 154, Part A, pp. 247–262.

Kitaoka, A., Gyoba, J., Kawabata, H. & Sakurai, K. (2001) Perceptual continuation and depth in visual phantoms can be explained by perceptual transparency. *Perception, 30*, 959–968.

Klymenko, V., & Weisstein, N. (1986) Spatial frequency differences can determine figure-ground organization. *Journal of Experimental Psychology, Human Perception & Performance, 12*, 324–330.

May, J. G., Brown, J. M., & Roberts, S. (1999) Afterimages, grating induction and illusory phantoms. *Vision Research, 39*, 3025–3031.

McCourt, M. E. (1982) A spatial frequency dependent grating-induction effect. *Vision Research, 22*, 119–134.

McCourt, M. E. (1990) Disappearance of grating induction at scotopic luminance. *Vision Research, 30*, 431–437.

McCourt, M. E. (1994) Grating induction: A new explanation for stationary phantom gratings. *Vision Research, 34*, 1609–1617.

Meng, M., Ferneyhough, E., & Tong, F. (2007) Dynamics of perceptual filling-in of visual phantoms revealed by binocular rivalry. *Journal of Vision*, 7, 1–15.

Meng, M., Remus, D. A., & Tong, F. (2005) Filling-in of visual phantoms in the human brain. *Nature Neuroscience, 8*, 1248–1254.

Metelli, F. (1974) The perception of transparency. *Scientific American, 230*, 90–98.

Mulvanny, P., Macarthur, R., & Sekuler, R. (1982) Thresholds for seeing visual phantoms and moving gratings. *Perception, 11*, 35–46.

Oyama, T., & Nakahara, J. (1960) The effect of lightness, hue and area upon the apparent transparency. *Japanese Journal of Psychology, 31*, 35–48.

Rosenbach, O. (1902) Zur Lehre von den Urtheilstäuschungen. *Zeitschrift für Psychologie, 29*, 434–448

Sakurai, K., & Gyoba, J. (1985) Optimal occluder luminance for seeing stationary visual phantoms. *Vision Research, 25*, 1735–1740.

Spillmann, L., Lehar, S., Stromeyer, M., & Wertheimer, M. (2006) *Laws of Seeing*. (Translation of W. Metzger's *Gesetze des Sehens*, 1st edition). Cambridge, MA: MIT Press.

Sugita, Y (1999) Grouping of image fragments in primary visual cortex. *Nature, 401*, 269–272.

Tynan, P., & Sekuler, R. (1975) Moving visual phantoms: A new contour completion effect. *Science, 188*, 951–952.

Weisstein, N. (1970) Neural symbolic activity: A psychophysical measure. *Science, 168*, 1489–1499.

Weisstein, N. (1973) Beyond the yellow-Volkswagen detector and the grandmother cell: A general strategy for the exploration of operations in human pattern recognition. In R. L. Solso (Ed.), *Contemporary Issues in Cognitive Psychology: Loyola Symposium*. Washington: V.H. Winston & Sons, pp. 17–51.

Weisstein, N., & Harris, C. S. (1974) Visual detection of line segments: An object superiority effect. *Science, 186*, 752–755.

Weisstein, N., & Harris, C. S. (1980) Masking and the unmasking of distributed representations in the visual system. In C. S. Harris (Ed.), *Visual Coding and Adaptability*. Hillsdale, NJ: Lawrence Erlbaum, pp. 317–364.

Weisstein, N., & Wong, W. (1986) Figure-ground organization and the spatial and temporal responses of the visual system. In E. C. Schwap & H. C. Nusbaum, (Eds) *Pattern Recognition by Human and Machines, Vol. 2, Visual Perception*. New York: Academic Press.

Weisstein, N., Maguire, W., & Berbaum, K. (1977). A phantom-motion aftereffect. *Science, 198*, 955–958.

Weisstein, N., Maguire, W., & Williams M. C. (1982) The effect of perceived depth on phantoms and the phantom motion aftereffect. In J. Beck (Ed.), *Organization and Representation in Perception*. Hillsdale, NJ: Lawrence Erlbaum, pp. 235–249.

Weisstein, N., Montalvo, F., & Ozog, G. (1972) Differential adaptation to gratings blocked by cubes and gratings blocked by hexagons: A test of the neural symbolic activity hypothesis. *Psychonomic Science, 27*, 89–92.

5

THE MOTION-INDUCED CONTOUR REVISITED

Gennady Erlikhman and Gideon P. Caplovitz
University of Nevada Reno

Introduction

In 1980, at the annual meeting of the Optical Society of America, Victor Klymenko and Naomi Weisstein presented a new visual phenomenon: an "Illusory contour produced by rotation-in-depth" (Klymenko and Weisstein 1980). The basic effect that they later called the "Motion-Induced Contour" is a vertically oriented illusory contour that is perceived when an outline projection of a rectangular solid is rotated in depth. The contour corresponds to an edge of the rectangular solid that would be there if in fact the solid object were present. This initial presentation was followed by four empirical papers describing a total of 16 experiments and 84 experimental conditions that together comprehensively explored the necessary and sufficient conditions for the illusory contour to manifest (Klymenko and Weisstein 1981, 1983, 1984; Klymenko et al. 1987). Nestled at the intersection of Gaetano Kanizsa's Illusory Contours and Hans Wallach's Kinetic Depth Effect, these empirical papers with titles that took us from "The edge of an event" (Klymenko and Weisstein 1983) to "The razor's edge" (Klymenko and Weisstein 1984) came at a time of an intense collective effort to understand of how representations of three dimensional (3D) objects were constructed from the 2D images projected on the retinae. Despite the high quality of these studies and the highly significant constraint the Motion-Induced Contour places on computational and neural models of object perception, the papers have only been cited a few dozen times in the literature and have not received the attention they perhaps deserve. In this chapter, we seek to shine a spotlight on these papers and celebrate the Motion-Induced Contour, not only for its theoretical impact but also for its phenomenological beauty that represents pure candy to any perceptual scientist. Here we describe and provide figural and video demonstrations of the Motion-Induced Contour phenomenon. We provide a review and summary of Klymenko and Weisstein's empirical findings and conclusions regarding the conditions under which the illusory contour is observed. We then discuss more recent phenomenological work on dynamic illusory contours highlighting how and why the Motion-Induced Contour is so special. Finally, we present some novel demonstrations of Motion-Induced Contours that were directly inspired by reading these classic papers in preparation for writing this chapter. These demonstrations both challenge and extend the conclusions of Klymenko and Weisstein, lay the groundwork for future empirical study, and further celebrate the lasting impacts of their work.

The Motion-Induced Contour

Describing novel perceptual phenomena and how they arise is not a trivial matter. Indeed, the first published description of the Motion-Induced Contour is quite a mouthful:

> An outline projection of a rectangular solid object, rotating in depth, produces a moving rotating in depth illusory contour across the empty region where an edge would normally be located even though it is not physically present.
>
> *(Klymenko and Weisstein 1981)*

As has always been the case for the vision scientist, a picture is worth a thousand words and a figure like that shown in Figure 5.1a, illustrating the inducing stimulus appeared on the same page as this initial description. Here we leverage the advantage of the digital-age vision scientist and provide a demonstration video[1] (Video 1) to allow you the reader to see the Motion-Induced Contour for yourself. The video depicts the wire-outline of a cube that is tilted toward the viewer and is rotating about the axis of tilt. The vertical edge closest to the observer is omitted. As should be readily apparent in the video, that moving edge, although not physically present, can be seen while the cube appears to rotate in depth. As was also noted in the original paper, flat, rigid, illusory surfaces are also perceived between the visible edges, meeting and forming the illusory edge in the middle.

Contemporary work on illusory contours often either simply described the phenomenology (e.g. Coren 1972; Ramachandran 1985) or else relied on subjective ratings of contour clarity as a method of quantifying illusory strength (e.g., Dumais and Bradley 1976). In all of their papers on the Motion-Induced Contour, Klymenko and Weisstein used the method of relative magnitude estimation by having subjects rate the perceived illusory contour strength relative to some comparison stimulus. Sometimes, the comparison stimulus was a rotating 3D object with real contours and illusory contour strength ratings were given

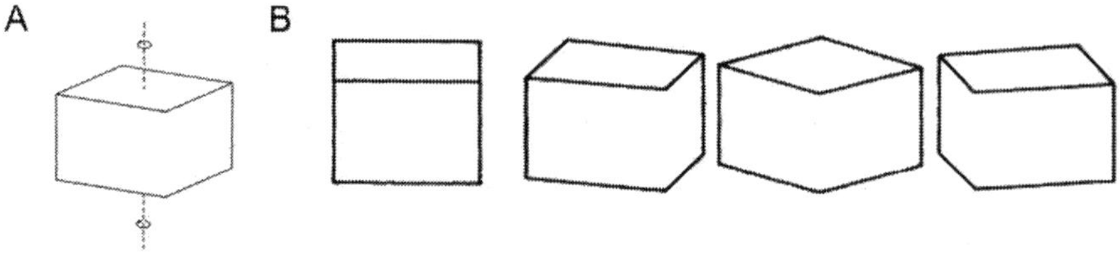

FIGURE 5.1

A. A wire cube, slightly tilted toward the observer rotates about the axis indicated by the dashed line. B. A sequence of frames from the stimulus in panel A and seen in Video 1. In the video, the cube is seen to rotate back and forth about a vertical axis. In the three rightmost images above, one of the edges of the cube is missing and is not visible. When the cube is set in motion, however, an illusory edge is seen in that position. This is the Motion-Induced Contour. [Figures 1 and 2 from Klymenko and Weisstein (1981). Reprinted by permission of SAGE Publications, Ltd.]

relative to this real contour for a simultaneously presented experimental stimulus (Klyemnko and Weisstein 1981, Experiment 1). In a majority of their experiments, multiple stimuli (up to five at a time) would be simultaneously presented within a circular array. The observers would be cued to first rate the strength of a contour within one of the five stimuli and then rate the relative strength of the others in the array (Klymenko and Weisstein 1983, 1984; Klymenko et al. 1987). By systematically varying the form and motion structure of the individual stimuli, they were able to quantify the relative strength of the Motion-Induced Contour under a wide range of stimulus conditions.

The primary goal of Klymenko and Weisstein's four papers on Motion-Induced Contours was to identify the necessary and sufficient conditions for perceiving the contour. In their 1981 paper, in addition to 3D rotation, they systematically tested various object transformations including a range of flicker rates, dichoptic presentation, misalignment, lateral translation (i.e., motion of the stimulus, but not in depth), and partial presentation of the objects (i.e. showing only a subset of the edges seen in Figure 5.1). Mean ratings of contour strength for four tested flicker frequencies (3.33, 5, 10, and 20 Hz) and for slow, medium, and fast lateral translations were no different from ratings of stationary displays: Motion-Induced Contour perception requires more than luminance change over time or 2D motion signals. When the object was rotating in depth, illusory strength magnitudes were two to three times stronger. Similarly large strength ratings were observed when the object was rotating about its axis of tilt (passing through the middle of the cube, which was slightly tilted forward) or about an axis coincident with the illusory contours (Experiment 2). In these cases, the illusory contour remained stationary on the screen while the cube rotated around it, indicating that motion of the contour itself was not necessary for the effect. Thus a key empirical finding from the 1981 paper was that Motion-Induced Contours only arise when the inducing contours are rotating in depth, giving rise to three dimensional structure (i.e., Kinetic Depth Effect: Wallach and O'Connell 1953) and are not visible when the visible parts of the figure are translating laterally, rotating in the 2D plane or are flickering (Klymenko and Weisstein 1981). This feature distinguishes Motion-Induced Contours from other types of "kinetic illusory contours" such as illusory phantoms and flicker induced contours, that were described extensively during this time period in a number of studies from different groups (e.g. Bradley and Lee 1982; Braunstein and Andersen 1981; Genter and Weisstein 1981; Kaplan 1969; Tynan and Sekular 1975; Weisstein et al. 1977).

A second finding from the 1981 paper, a finding that greatly influenced the choice of stimuli in subsequent experiments, was that it was not necessary to have a complete rectangular volumetric object to give rise to the Motion-Induced Contour. The Motion-Induced Contour is equally strong even when two or four edges were removed such that only two inducing vertices defined the stimulus (Figure 5.2 and Video 2A). As can be seen in Video 2A, simply having two vertices aligned to represent a dihedral (defined by two planar faces) object is sufficient to produce the Motion-Induced Contour which in this case connects the apices of each vertex, much like the spine of an open book. If, instead of taking the front two vertices of the virtual cube, the vertices are

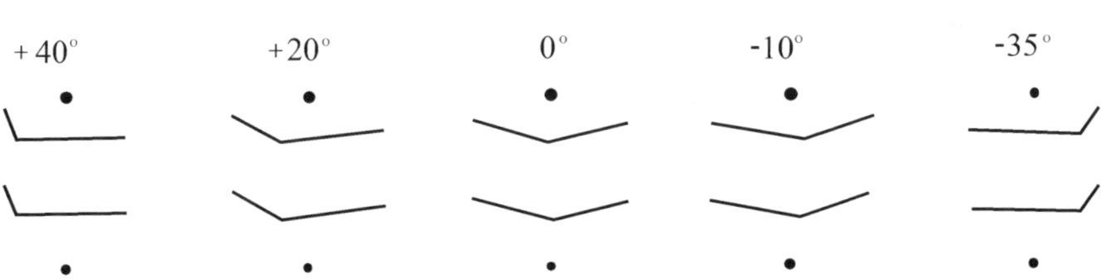

FIGURE 5.2 The cube from Figure 5.1, except with the "sides" and top edges removed, leaving only two "chevrons". Each panel shows a different view of the chevrons as they rotate about the axis of tilt.

As with the more complete figure, no edge is seen connecting the two apices of the chevrons in any single image. However, when shown in sequence as an animation of a chevron rotated in depth, an illusory edge is seen connecting the apices and illusory surfaces appears between the edges. [Figure 5A from Klymenko and Weisstein (1983), The edge of an event: Invariants of a moving illusory contour. Perception & Psychophysics, *34(2), pp. 140–148, reprinted with permission by Springer Nature].*

taken from the back (Video 2B), an illusory contour is seen as well and may again appear to be a front-edge of a dihedral surface that is tilted up rather than down as in Video 2A. However, this perspective is ambiguous and the stimulus can also look like a book open towards the observer. We note that the open-towards/open-away perspective is ambiguous, and with some attentive-effort, the stimuli in 2A and 2B can each be seen either way. Compare them side-by-side in Video 2C.

An illusory contour can still be seen even if only one of the line segments making up each vertex is shown (Klymenko and Weisstein 1984; Video 3). One can see both kinds of edges in Videos 2A and 2B: the dihedral edge is formed at the intersection of two surfaces between the apices (i.e., the spine of the book). A "monohedral" or "razor" edge can also be seen connecting the endpoints of the line segments on the left and right sides of the shape (i.e. the edges of the book). The dihedral edge is not necessary for the perception of monohedral edges: in Video 3, only the right side of the shapes from Video 2A are shown, resulting in a percept of "half of a book", or, alternatively, a closed book. Here, illusory contours are seen joining the endpoints of each line segment and an illusory surface may be perceived between the two lines. In rating experiments, the illusory contour forming the monohedral edge is not as strong as that of dihedral edges. One of the take home messages of the 1984 paper is that unlike the dihedral edge, the monohedral edge is equally strong during rotation in depth and planar motion, suggesting that monohedral edges may arise from a different mechanism than Motion-Induced Contours.

In subsequent work, Klymenko and Weisstein systematically investigated what aspects of motion and depth were necessary for perceiving Motion-Induced Contours: whether rotation alone, rotation in depth, other motion in depth (i.e. looming), or other types of depth information (i.e. disparity cues) were necessary or sufficient to produce the effect. Klymenko and Weisstein compared the perceived illusory contour strength of two pairs of vertices or "chevrons" (Figure 5.2) undergoing various transformations (Klymenko and Weisstein 1983; Experiment 1) The chevrons would either rotate in depth, translate laterally, flicker, or be stationary. A critical new condition was added in which the chevrons were also rotating in depth, but in which they had a different 3D structure. Recall that the original stimulus was defined as the sequence of frames produced from the parallel projection of a cube that was rotating in depth with only four of its edges visible, making the top and bottom vertices or chevrons. Although each individual image had no depth information, in the animation sequence, the apices of the chevrons appeared closer to the observer while the ends appeared farther away. In the new stimulus, the sequence of 2D images produced by rotation of the chevrons were generated not from chevrons arranged in depth, but from chevrons that

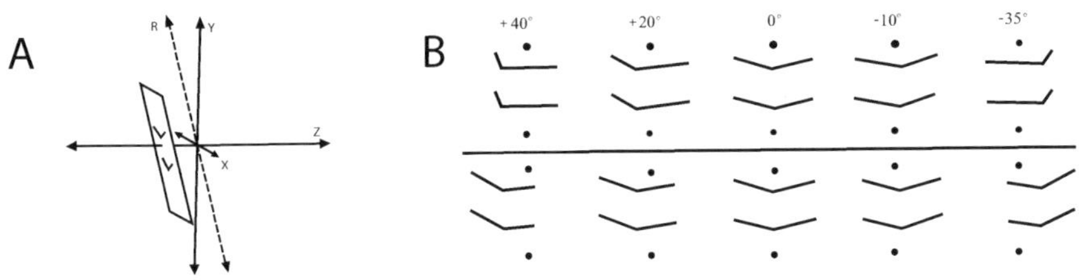

FIGURE 5.3

A. Tilted chevrons from Klymenko and Weisstein (1983). The chevrons are coplanar. The entire plane rotates about the vertical Y axis. B. Top row—sequence of frames from the rotation of chevrons derived from the edges of a rotating cube. Bottom row—sequence of frames from the rotations of chevrons that are coplanar as in panel A. [Figures 5 and 6 reprinted by permission from Springer Nature.]

were parallel to each other on a single plane which itself rotated about the vertical axis. This is depicted in Figure 5.3. Here, the apices and ends of the chevron lie on the same depth plane. The sequence of transformations produced by the rotation of this plane is shown in the bottom row of Figure 5.3. Contrast this to the sequence of transformations produced by the original stimulus derived from a cube's rotation, shown in the top row: they are extremely similar. Despite the relatively minor differences between the two sequences, the resulting percepts are dramatically different: a Motion-Induced Contour and a sense of rotation in depth is seen only for the top row of frames, created from a rotating cube. Illusory contour magnitude for the new stimulus was much smaller, similar to that of flickering, translating, and stationary stimuli which also did not produce Motion-Induced Contours.

This experiment demonstrated that rotation in depth alone was insufficient to produce illusory contours. While insufficient, some rotational motion may be necessary. Klymenko, Weisstein, and Ralston (1987) tested whether other forms of depth information such as stereo presentation (Experiment 1), perspectival cues (i.e. polar projection, Experiment 2), looming (Experiments 3a and 3b), and rotation + change in size (Experiment 4). Adding disparity information to the rotating stimulus did not significantly increase the strength of the illusory contour. Contour strength was rated to be much weaker for other cues or when the object loomed, but did not rotate. Rotation, therefore, provides critical information about the rigid 3D structure of the object that is not available from other kinds of transformations or cues. Furthermore, Motion-Induced Contours were seen even if the upper and lower portions were presented dichoptically, suggesting that the illusory contour was formed only after information from each eye was combined. Other figural cues are

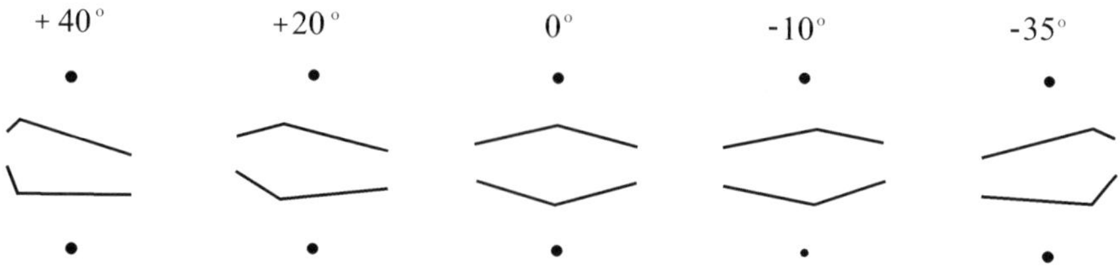

FIGURE 5.4 A sequence of frames of an "oblique" Motion-Induced Contour, similar to Figure 5.2, except with the top chevron angled upward.

The resulting illusory contour appears angled toward the observer, connecting the two chevron apices. An animated version can be seen in Video 5. [Figure 8A from Klymenko and Weisstein (1983). The edge of an event: Invariants of a moving illusory contour. Perception & Psychophysics, *34(2), pp. 140–148, reprinted with permission by Springer Nature].*

also not necessary for the perception of Motion-Induced Contours: if one of the vertices is flipped upside down so that the edges are no longer parallel, illusory contours are still seen provided that they still describe rigid 3D motion (Klymenko and Weisstein 1983, Experiments 3 and 4; Video 4). Such a display can be created by moving the vertex point from the upper chevron upward or, as originally described in their paper, by drawing two downward diagonal edges from the upper vertex that lie on the dihedral surfaces as indicated by the dashed lines in Figure 4. Illusory contours are also seen in apparent motion sequences and do not depend on the smooth rotation of the object (Experiment 3, Video 5).

Over the course of four papers, Klymenko and Weisstein meticulously characterized the properties of a novel visual illusion—the Motion-Induced Contour—in which one sees kinetic depth and a rigid dihedral edge. Klymenko and Weisstein concluded that there was no single static cue or motion type that on their own always produced Motion-Induced Contours. Rather, the sequence of 2D images must specify the "structural invariant" of a rigid dihedral edge. A contemporary take on this phenomenon might refer to a bias or prior for slow and smooth rigid motion as opposed to an interpretation of the visible edges as deforming or changing in size. This does not explain, however, why an illusory contour is seen between the apices—it only describes why a sequence of 2D images might have a particular 3D interpretation. To begin to address the question of how Motion-Induced Contours might come about (a question not tackled in Klymenko and Weisstein's papers), it is instructive to compare Motion-Induced Contours to other types of kinetic and volumetric illusory contours and surfaces.

Relationship to Other Perceptual Phenomena

Motion-Induced Contours share several features with other well-known phenomena at the intersection of form and motion perception. One of the characteristic features of Motion-Induced Contours is that they are only seen when the vertices are set in motion, and not in any static 2D views (Klymenko and Weisstein 1981). Sequences of 2D images can produce percepts of 3D forms in other ways. In the stereokinetic effect (Musatti 1924; Proffitt et al. 1992), depth is seen as the result of object rotation in the picture plane. Motion-Induced Contours are likely the result of a different process since neither they nor 3D structure are seen when the vertices are rotated in the picture plane or on a plane parallel to the axis of tilt (Klymenko and Weisstein 1983). 3D form from successive 2D images can also be seen in the kinetic depth effect and structure-from-motion (Ullman 1979; Wallach and O'Connell 1953). As for Motion-Induced Contours, rigidity (Johansson 1973; Ullman 1979, 1984a, 1984b) and slow and smooth velocities (Weiss and Adelson 1998; Weiss et al. 2002) may be important constraints for recovering structure. Additionally, as in structure-from-motion, replacing the vertices in Klymenko and Weisstein's display with a smooth contour weakens the 3D interpretation (Braunstein and Andersen 1984). However, structure-from-motion can also be seen for non-rigid forms (Braunstein and Andersen 1984; Erlikhman et al. 2014; Jain and Zaidi 2011; see Vezzani et al. 2014 for a review) perhaps distinguishing it from Motion-Induced Contours.

Illusory contours that are only seen through motion also occur in random dot kinematograms when, across two frames, a region of black and white dots is displaced (Braddick 1974; Julesz 1971; Petersik et al. 1978), during kinetic occlusion, when one surface of the same color as the background gradually occludes another or when textured surface slides behind another similarly textured surface (Bradley and Lee 1982; Gibson 1966, 1979; Gibson et al. 1969; Kaplan 1969; Kellman and Cohen 1984; Michotte et al., 1964), in spatiotemporal boundary formation in which no inducing edges are seen on any single frame, but an illusory contour appears through the sequential transformations of individual texture elements (Erlikhman and Kellman 2016; Hine 1987; Prazdny 1986; Shipley and Kellman 1994), and in spatiotemporal form integration in which the inducing elements that support the illusory contour are only gradually revealed over time (Kojo et al. 1993; McCarthy et al. 2015; Palmer et al. 2006). Motion-Induced Contours are distinct from most of these phenomena in that no part of the display is gradually occluded and disoccluded as the vertices rotate. The cue to structure in Motion-Induced Contours is not texture accretion and deletion. This may be a critical, distinguishing feature because, to our knowledge, structure-from-motion only produces straight illusory contours at the boundaries of an object, e.g. the sides of a rotating cylinder. In contrast, the Motion-Induced Contour is a dihedral edge, separating two surfaces of the object and not the object from the background.

The fact that Motion-Induced Contours do not specify an occlusion boundary differentiates them from other illusory contours which can be readily seen in static displays such as in Kanizsa figures (Kanizsa 1955, 1979; Schumann 1904). In such static displays, contours are seen to "emanate" from line endings or terminators (Field et al. 1993; Kellman and Shipley 1991; von der Heydt et al. 1984), connect two line endings in a smooth path (Kellman and Shipley 1991), or appear between coincidentally aligned line endings (Ehrenstein 1941; Gibson 1950; Gillam 1987; Gillam and Chan 2002; Vogels and Orban 1987). In the Kanizsa triangle, for example, an illusory triangle appears to be in front of three circles (Kanizsa 1979). L-junctions in the image act as occlusion cues indicating a depth ordering (Guzman 1969; Kellman and Shipley 1991; Rubin 2001; Shipley and Kellman 1990). Motion-Induced Contours are unusual in that an illusory contour is perceived between two vertices, extending in a direction that is not along the length of either of the edges forming the vertex. This may be related to why the dihedral edge is not the boundary between a figure and background or between two surfaces belonging to different objects. Rather, it is the edge connecting two surfaces of a single, rigid object. It is worth noting that even in displays with volumetric illusory contours, the illusory edges of the illusory volumes are also only seen at the meeting of one surface and another, that is, at an occlusion boundary (e.g. Carman and Welch 1992; Kellman et al. 2005b; Tse 1998, 1999), and while there are other examples of 3D interpolation in which an illusory surface is perceived between two converging planes (e.g. Kellman et al. 2005a), the interpolated region in those displays is a smooth, curved surface as opposed to two surfaces meeting at an edge. This property of the Motion-Induced Contour—that it is not visible in static views and that it is not an occlusion boundary—highlights the fact that the dihedral edge is derived from motion information and not from any static 2D cues to form such as luminance contrast or contour junctions.

Demonstration Videos of Novel Variants

In writing this chapter, it was our goal to both highlight Klymenko and Weisstein's work on Motion-Induced Contours and also to re-emphasize their importance and potential as a tool for better understanding the mechanisms that underlie 3D form and motion perception. To this latter end, we have created a series of demonstration videos that recreate some of the original stimuli that lead to Motion-Induced Contours. Inspired by viewing these demonstrations, we created a series of novel variants of Klymenko and Weisstein's older displays. While some of these variants lead to counterintuitive observations that potentially challenge their conclusions, each is designed to illustrate that Klymenko and Weisstein's work on Motion-Induced Contours is far from finished and represents fertile ground from which new lines of empirical research can grow.

Inspired by their papers, we present several novel effects relating to Motion-Induced Contours in Videos 6–15. First, we considered configural effects of surface completion and surface relationships on the perception of monohedral and dihedral Motion-Induced Contours.

Video 6: By adding circles around the edge terminators, we can block the formation of monohedral edges at the sides of the display in Video 2A. This weakens the perception of surfaces between the chevrons, but the dihedral Motion-Induced Contour can still be seen.

Video 7A: Vertical lines added to the vertices of the chevron stimulus from Video 2A, converting the L-junctions into Y-junctions. If surfaces are necessary for the perception of the dihedral edge, additional global cues suggesting that the chevrons belong to objects above and below where the dihedral edge should appear may disrupt surface formation and the perception of the dihedral edge. A Motion-Induced Contour can still be seen, although perhaps it is a little bit weaker.

Video 7B: Same as Video 7A except the vertical lines are added to the edge terminators, converting them into L-junctions. The monohedral edges are no longer seen, but the dihedral edge is still visible. Of note is that the dihedral edge can be seen even if the stimulus is perceived as two solids, one on top and one on bottom, in which case the edge appears to span empty space.

Video 7C: Closure is added to Video 7B to strengthen the global cue by creating two copies of the stimulus from Video 1 one above the other. A dihedral edge is seen in each of the two objects, but an additional edge is still seen between them. This occurs despite the boundary ownership cues indicating that the edges forming vertices where the illusory contour appears belong to another surface, almost as though space itself has been warped in 3D—as though the pixels in the middle of the display appear closer to us, even though there is nothing in that space specifying a surface. As in Video 7B, the monohedral edges, however, are not visible.

Video 8: The perception of surfaces can also be eliminated by taking two opposite edges from Video 2A. Here two surfaces can be inferred, such that they are organized much like a laptop that is opened away from the observer, with one surface upright and the other receding in depth. However, no illusory surfaces are seen, yet a crisp vertical Motion-Induced Contour is observed. This contour represents a monohedral edge as belonging to only one of the inferred surfaces (the one of the right, if the edges are taken to be oriented away from the observer, and the one on the left if the edges are seen as opening toward the observer). The dihedral edge corresponding to the "hinge" of the laptop is not seen.

Next, we introduce several new manipulations of the original stimulus to examine whether properties that are known to affect the representation of 3D form and perceived strength of other forms of illusory contours also modify Motion-Induced Contours.

Video 9A: Static illusory contours become stronger as function of the ratio of the specified (i.e. visible) edge to total edge length (i.e. visible + illusory; Shipley and Kellman 1992). In this context the support ratio of the Motion-Induced Contour is nearly zero, as only the terminators are explicitly defined. In this demonstration, we manipulated the length of the vertices forming each chevron and in an analogous fashion to the static-support ratio found that as the length of the edges decreased, the perceived strength of the Motion-Induced Contour also diminished until it was no longer visible. Monohedral edges connecting the endpoints of the chevrons are also weaker, perhaps decreasing in strength at the same rate as the dihedral edge. In contrast to static 2D contours, the dihedral edge was still quite strong even when the total length of the edge segments was much smaller than the distance between the two vertices (which are constant across the three examples).

Video 9B: Making the lines thinner, has little effect on the configural relationships observed in Video 9A further suggesting that the classic support ratio does not apply to the Motion-Induced Contour.

Video 9C: By putting small gaps in the chevrons, the dihedral edge percept is weakened, but now monohedral edges and moving phantoms may be seen. In this display, the length of the edge segments making the corner is the same as the leftmost stimulus in Video 9B whereas the total length of the edges is the same as the original stimulus in Video 2A. Probing the influences of additional configural cues may provide a fruitful domain for future research.

Video 10: In this video, instead of removing parts of the chevrons toward each end as in Video 9, the apex and surrounded edges are removed. Although monohedral edged surfaces are perceived, no dihedral Motion-Induced Contour is seen. In this case, as with other examples of 3D illusory contours, corners are not interpolated if the vertices are occluded or missing (Kellman et al. 2005a). We also note the emergence of moving phantoms, a phenomenon that Weisstein studied extensively in other work (Weisstein et al. 1977).

Video 11: Despite the representation of 3D form, Motion-Induced Contours are not perceived in classic structure-from-motion displays.

Video 12: Nor are Motion-Induced Contours seen in other illusory 3D volumetric figures. Here, the surfaces of a cube are specified by other elements as in Tse (1999): a black rotating cube on a black background has cylindrical white spikes emanating from the center of each of its faces.

Video 13A: If the entire shape is rounded, no illusory contour or surfaces are seen. The sharp corner appears to be necessary.

Video 13B: However, if the rounding is restricted to the vertex, then a curved illusory surface can be perceived perhaps emanating from the monohedral edges which are visible in this video, but weaker than in Video 13A. The degree of curvature located along the apices of the rounded chevrons may influence whether or not and how strongly a Motion-Induced Contour is perceived. If the video is made small, the corresponding region of high curvature begins to approximate a corner (Caplovitz et al. 2006) and a somewhat weak Motion-Induced Contour may be seen.

Finally, we note that of the many possible 3D interpretations of the sequence of 2D images in Figure 5.1 and the top row of Figure 5.3B, a rigid figure is preferred. Alternative interpretations could be of figures that are changing in size, shape, and velocity frame to frame. There is a long history of studies supporting the perceptual priority of rigid motion, particularly in the construction of 3D structure and motion from the spatiotemporal sequence of 2D images that arise on the retinae (Jansson and Johansson 1973; Ullman 1984b). Indeed, one of the primary conclusions that Klymenko and Weisstein derive from the work is the representational necessity of rigid 3D-rotation for the formation of the dihedral Motion-Induced Contours (Klymenko and Weisstein 1983). However, there are kinds of dynamic stimuli in which non-rigid illusory contours can be seen (e.g. Erlikhman et al. 2014; Jain and Zaidi 2011). Here, we attempt to create a non-rigid version of Klymenko and Weisstein's display.

Video 14A: A "twisting" illusory contour. The same as Video 1, except the top and bottom of the shape are set to rotate in opposite directions. The entire shape may appear like a sponge that is being twisted or wrung.

Video 14B: A chevron version of Video 14A. A twisting contour can still be seen.

Video 15: Taking the bottom chevron from Video 2A and the top chevron from Video 2B (i.e. a front and back edge of a cube). If the chevrons are perceived in this way (bottom chevron pointing toward the observer and top chevron pointing away), then no Motion-Induced Contour is seen between the apices. Such a contour would be tilted in depth. However, if both chevrons are perceived to be pointing in the same direction (either toward or away from the observer), then a twisting illusory contour can be seen as in Video 14B. Seeing it this way can be facilitated by covering the bottom chevron, and trying to see the upper chevron as pointing towards you and then uncovering the bottom chevron. One fascinating aspect of this phenomenon is that, with practice you can voluntarily switch between the two views, then the perception of an illusory contour is placed under voluntary control, depending on the organization of the scene (see also Tse et al. 2013).

Conclusion

Through their discovery of and empirical work describing the Motion-Induced Contour, Klymenko and Weisstein blazed a trail from "The edge of an event" to "The razor's edge", that represents a journey well travelled by the perceptual scientist grounded in Gestalt Psychology. What are these phantasmic contours, why do they manifest, and what do they tell us about the nature of information the brain uses to construct 3D representations of objects in motion? Despite the great progress Kylmenko and Weisstein made in answering such questions, our new demonstrations highlight the fact that much work remains to fully understand the nature of Motion-Induced Contours and how they fit in with modern conceptualizations of 2D and 3D figural and motion perception. Moreover, during the past decades since the Motion-Induced Contour was first described, great strides have been made in neuroimaging technologies that have led the field to more directly relate such phenomenological work to underlying brain structure and neural mechanisms. It is our hope that the Motion-Induced Contour and the legacy of Klymenko and Weisstein's seminal work will live on to serve as a valuable tool to both better understand the phenomenological nature of our visual experiences and their neuronal bases.

Note

1 To view all videos referred to in Chapter 5 go to https://research.franklin.uga.edu/visual-perception-laboratory/chapter-5

References

Braddick, O. (1974). A short-range process in apparent motion. *Vision Research, 14*, 519–527.

Bradley, D. R., & Lee, K. (1982). Animated subjective contours. *Perception & Psychophysics, 32*(4), 393–395.

Braunstein, M. L., & Anderson, G. J. (1981). Velocity gradients and relative depth perception. *Perception & Psychophysics, 29*, 145–155.

Braunstein, M. L., & Anderson, G. J. (1984). A counterexample to the rigidity assumption in the visual perception of structure from motion. *Perception, 13*, 213–217.

Caplovitz, G. P., Hsieh, P.-J., & Tse, P. U. (2006). Mechanisms underlying the perceived angular velocity of a rigidly rotating object. *Vision Research, 46*, 2877–2893.

Carman, G. J., & Welch, L. (1992). Three-dimensional illusory contours and surfaces. *Nature, 360*, 585–587.

Coren, S. (1972). Subjective contours and apparent depth. *Psychological Review, 79*, 359–367.

Dumais, S. T., & Bradley, D. R. (1976). The effects of illumination level and retinal size on the apaprent strength of subjective contours. *Perception & Psychophysics, 19*(4), 339–345.

Ehrnstein, W. (1941). Uber Abwandlungen der L. Hermannschen Helligkeiterscheinung. *Zeitscrhift für Psychologie, 150*, 83–91.

Erlikhman, G., & Kellman, P. J. (2016). From flashes to edges to objects: Recovery of local edge fragments initiates spatiotemporal boundary formation. *Frontiers in Psychology, 7*, 910.

Erlikhman, G., Xing, Y. Z., & Kellman, P. J. (2014). Non-rigid illusory contours and global shape transformations defined by spatiotemporal boundary formation. *Frontiers in Human Neuroscience, 8*, 1–13.

Field, D. J., Hayes, A., & Hess, R. F. (1993). Contour integration by the human visual system: Evidence for a local "association field". *Vision Research, 33*(2), 173–193.

Genter, C. R. II., & Weisstein, N. (1981). Flickering phantoms: A motion illusion without motion. *Vision Research, 21*, 963–966.

Gibson, J. J. (1950). *The Perception of the Visual World*. Oxford: Houghton Mifflin.

Gibson, J. J. (1966). *The Senses Considered as Perceptual Systems*. Oxford: Houghton Mifflin.

Gibson, J. J. (1979). *The Ecological Approach to Visual Perception*. Boston, MA: Houghton, Mifflin, and Co.

Gibson, J. J., Kaplan, G. A., Reynolds, H. N., & Wheeler, K. (1969). The change from visible to invisible: A study of optical transitions. *Perception & Psychophysics, 5*, 113–116.

Gillam, B. (1987). Perceptual grouping and subjective contours. In S. Petry & G. E. Meyer (Eds) *The Perception of Illusory Contours* (pp. 268–273). New York, NY: Springer-Verlag.

Gillam, B., & Chan, W. M. (2002). Grouping has a negative effect on both subjective contours and perceived occlusion at T-junctions. *Psychological Science, 13*(3), 279–283.

Guzman, A. (1969). Decomposition of a visual scene into three-dimensional bodies. In A. Grasselli (Ed.) *Automatic Interpretation and Classification of Images* (pp. 243–276). New York, NY: Academic Press.

Hine, T. (1987). Subjective contours produced purely by dynamic occlusion of sparse-points array. *Bulletin of the Psychonomic Society, 25*(3), 182–184.

Jain, A., & Zaidi, Q. (2011). Discerning nonrigid 3D shapes from motion cues. *Proceedings of the National Academy of Sciences of the USA, 108*(4), 1663–1668.

Jansson, G., & Johansson, G. (1973). Visual perception of bending motion. *Perception, 2*, 321–326.

Johansson, G. (1973). Visual perception of biological motion and a model for its analysis. *Perception & Psychophysics, 14*, 201–211.

Julesz, B. (1971). *Foundations of Cyclopean Perception*. Chicago, IL: University of Chicago Press.

Kanizsa, G. (1955). Margini quasi-percettivi in campi con stimolazione omogenea. *Rivista di Psicologia, 49*, 3–19.

Kanizsa, G. (1979). *Organization in vision: Essays on Gestalt Perception*. New York, NY: Praeger.

Kaplan, G. A. (1969). Kinetic disruption of optical texture: The perception of depth at an edge. *Perception & Psychophysics, 6*, 193–198.

Kellman, P. J., & Cohen, M. H. (1984). Kinetic subjective contours. *Perception & Psychophysics, 35*(3), 237–244.

Kellman, P. J., & Shipley, T. F. (1991). A theory of visual interpolation in object perception. *Cognitive Psychology, 23*, 141–221.

Kellman, P. J., Garrigan, P., & Shipley, T. F. (2005a). Object interpolation in three dimensions. *Psychological Review, 112*(3), 586–609.

Kellman, P. J., Garrigan, P., Shipley, T. F., Yin, C., & Machado, L. (2005b). 3-D interpolation in object perception: Evidence from an objective performance paradigm. *Journal of Experimental Psychology: Human Perception and Performance, 31*, 558–583.

Klymenko, V., & Weisstein N. (1980). Illusory contour produced by rotation-in-depth. *Optics News, 6*(3), 61. Abstract.

Klymenko, V., & Weisstein N. (1981). The motion-induced contour. *Perception, 10*, 627–636.

Klymenko, V., & Weisstein N. (1983). The edge of an event: Invariants of a moving illusory contour. *Perception & Psychophysics, 34*(2), 140–148.

Klymenko, V., & Weisstein N. (1984). The razor's edge: A dichotomy between monohedral and dihedral edge perception. *Vision Research, 24*(9), 995–1002.

Klymenko, V., Weisstein N., & Ralston, J. V. (1987). Illusory contours, projective transformations, and kinetic shape perception. *Acta Psychologica, 64*, 229–243.

Kojo, I., Liinasuo, M., & Rovamo, J. (1993). Spatial and temporal properties of illusory figures. *Vision Research*, *33*, 897–901.

McCarthy, J. D., Strother, L., & Caplovitz, G. P. (2015). Spatiotemporal form integration: Sequentially presented inducers can lead to representations of stationary and rigidly rotating objects. *Attention, Perception, & Psychophysics*, 77(8), 2740–2754.

Michotte, A, Thinés, G., & Crabbé, G. (1964). *Les compléments amodaux des structures perceptives*. Louvain, Belgium: Studia Psychologica Louvain: Publications Universitaires.

Musatti, C. L. (1924). Sui denomeni stereocinetici. *Archivio Italiano di Psicologia*, *3*, 105–120.

Palmer, E. M., Kellman, P. J., & Shipley, T. F. (2006). A theory of dynamic occluded and illusory object perception. *Journal of Experimental Psychology: General*, *135*(4), 513–541.

Petersik, J. T., Hicks, K. I., & Pantle, A. J. (1978). Apparent movement of successively generated subjective figures. *Perception*, 7, 371–383.

Prazdny, K. (1986). Illusory contours from inducers defined solely by spatiotemporal correlation. *Perception & Psychophysics*, *39*(3), 175–178.

Proffitt, D. R., Rock, I., Hecht, H., & Schubert, J. (1992). Stereokinetic effect and its relation to the kinetic depth effect. *Journal of Experimental Psychology: Human Perception & Performance*, *18*, 3–21.

Ramachandran, V. S. (1985). Apparent motion of subjective surfaces. *Perception*, *14*, 127–134.

Rubin, N. (2001). The role of junctions in surface completion and contour matching. *Perception*, *30*, 339–366.

Schumann, F. (1904). Einige Beobachtungen über die Zusammenfassung von Gesichtseindrucken zu Einheiten. *Psychologische Studien*, *1*, 1–32.

Shipley, T. F., & Kellman, P. J. (1990). The role of discontinuities in the perception of subjective figures. *Perception & Psychophysics*, *48*(3), 259–270.

Shipley, T. F., & Kellman, P. J. (1992). Strength of visual interpolation depends on the ratio of physically specified to total edge length. *Perception & Psychophysics*, *52*(1), 97–106.

Shipley, T. F., & Kellman, P. J. (1994). Spatiotemporal boundary formation: Boundary, form, and motion perception from transformations of surface elements. *Journal of Experimental Psychology: General*, *123*(1), 3–20.

Tse, P. (1998). Illusory volumes from conformation. *Perception*, *27*, 977–992.

Tse, P. (1999). Volume completion. *Cognitive Psychology*, *39*, 37–68.

Tse, P. U., Reavis, E. A., Kohler, P. J., Caplovitz, G. P., & Wheatley T. P. (2013). How attention can alter appearances. In L. Albertazzi (Ed.) *The Wiley-Blackwell Handbook of Experimental Phenomenology: Visual Perception of Shape, Space and Appearance*. Chichester, UK: John Wiley & Sons, Ltd.

Tynan, P., & Sekuler, R. (1975). Moving visual phantoms: A new contour completion effect. *Science*, *188*, 951–952.

Ullman, S. (1979). *The Interpretation of Visual Motion*. Cambridge, MA: MIT Press.

Ullman, S. (1984a). Rigidity and misperceived motion. *Perception*, *13*, 219–220.

Ullman, S. (1984b). Maximizing rigidity: The incremental recovery of 3-D structure from rigid and non-rigid motion. *Perception*, *13*, 255–274.

Vezzani, S., Kramer, P., & Bressan, P. (2014) Stereokinetic effect, kinetic depth effect, and structure from motion. In J. Wagemans (Ed.) *The Oxford Handbook of Perceptual Organization*. Oxford: Oxford University Press.

Vogels, R., & Orban, G. A. (1987). Illusory contour orientation discrimination. *Vision Research*, 27(3), 453–467.

von der Heydt, R., Peterhans, E., & Baumgartner, G. (1984). Illusory contours and cortical neuron responses. *Science, 224*, 1260–1262.

Wallach, H., & O'Connell, D. N. (1953). The kinetic depth effect. *Journal of Experimental Psychology, 45*(4), 205–217.

Weiss, Y., & Adelson, E. H. (1998). Slow and smooth: A Bayesian theory for the combination of local motion signals in human vision. *MIT AI Lab Memo*, AI Memo 1624.

Weiss, Y., Simoncelli, E. P., & Adelson, E. H. (2002). Motion illusions as optimal percepts. *Nature Neuroscience, 5*(6), 598–604.

Weisstein, N., Maguire, W., & Berbaum, K. (1977). A phantom-motion aftereffect. *Science, 198*(4320), 955–958.

6

MATHEMATICAL, ARCHITECTURAL, AND FUNCTIONAL FOUNDATIONS OF VISUAL MASKING

Haluk Öğmen
University of Denver

Introduction

Visual masking refers to the reduced visibility of one stimulus, called the target, due to the presence of another stimulus, called the mask (Weisstein, 1972; Bachmann, 1994; Breitmeyer & Öğmen, 2000, 2006). When the mask is presented before (after) the target, masking is said to be forward (backward) masking. Weisstein started her 1968 *Psychological Review* paper (Weisstein, 1968) by considering three broad classes of explanations offered for visual masking: 1. confusion explanations, 2. substitution or overtake explanations, and 3. explanations using lateral inhibition.

The confusion hypothesis states that if the mask is similar to the target stimulus, it would make the target difficult to perceive, identify, or discriminate as an *individual* stimulus relative to the *entire* stimulus consisting of both the target and the mask. A prediction of this hypothesis is that, the closer in time the mask and the target are presented with respect to each other, the stronger the masking would be. Hence, if we were to plot some measure of target visibility as a function of the delay between the target and the mask (SOA: stimulus onset asynchrony; positive (negative) SOAs correspond to the case where the mask is presented after (before) the target), the maximum masking should occur at SOA=0 and the strength of masking should decrease as SOA becomes more negative or more positive. Hence, considering backward and forward masking separately (i.e., negative and positive SOAs, respectively) one obtains *monotonic* masking functions for each. Indeed, this type of masking-function is often observed and is called "Type-A" or "monotonic" masking functions (Bachmann, 1994; Breitmeyer & Öğmen, 2000, 2006). However, another type of masking function is the so-called "Type-B" or "non-monotonic" or "U-shaped" masking function, where the maxima of masking do not occur at SOA=0, but instead occur at positive and negative values of SOA, thereby producing non-monotonic forward and backward masking functions (Bachmann, 1994; Breitmeyer & Öğmen, 2000, 2006). Weisstein noted that this formulation of the confusion hypothesis can account for Type-A masking functions but not for Type-B masking functions.

The stimulus substitution (overtake) hypothesis states that the mask stimulus substitutes, or overtakes, the target stimulus somewhere along the retino-cortical pathways.[1] In this basic form, this hypothesis would predict Type-A, but not Type-B masking functions. The Averbach and Coriell (1961) version of this hypothesis states that substitution can take place in short-term memory. When the capacity of short-term memory is reached, in order to introduce a new item, a previously held item needs to be discarded, hence leading to the substitution of one item (target) by another item (mask). This version can predict Type-B masking functions because a mask stimulus that is presented *after* a target stimulus can substitute the target stimulus. This is because both the target and mask stimuli can be held in sensory memory for a few hundred milliseconds. The transfer from sensory memory to short-term memory or substitution in short-term memory can take place among any items in sensory or short-term memory irrespective of their temporal order of presentation. However, as noted by Weisstein, there are many shortcomings in this approach. First, Type-B masking can occur even when the entire stimulus is within the capacity limits of STM. Hence there is no *a priori* reason for substitution in STM. Furthermore, even if substitution occurs, there is no *a priori* systematic necessity dictating which stimulus should substitute which one. As a result, the outcome should be rather ad-hoc and erratic rather than following the systematic and lawful relations observed in visual masking. Second, even when capacity limits are exceeded, the transfer from sensory memory to STM is a flexible and task-dependent process and hence leading to subject- and task-dependent outcomes rather than systematic relations observed in the masking literature. Furthermore, Weisstein argued that this explanation can account for Type-A or Type-B functions but not both.

Finally, the third hypothesis considered by Weisstein was based on lateral inhibition. From a more contemporary perspective, one would not consider these three hypotheses at the same level, since the first two refer to higher-level processes whereas lateral inhibition is a more basic low-level neural mechanism. Nevertheless, neurophysiological studies in the 1950s and 1960s highlighted the role of lateral inhibition as a fundamental operation in neural processing and it was directly linked to perceptual phenomena, such as Mach bands (Ratliff, 1965). Since inhibition reduces or suppresses neural activity, and since in masking the target stimulus's visibility is reduced or suppressed, it was a natural quest to explore if lateral inhibition can explain visual masking. In fact, this is the path that Weisstein took in building her model of visual masking.

Mathematical Foundations

Before introducing Weisstein's model, let us first introduce some of the fundamental mathematical formulations of neural activity. Many of the building blocks of mathematical neuroscience were already laid out in 1950s and 1960s. An extensive review and comparison of these models can be found in Grossberg (1988). Some notable examples are the Hodgkin-Huxley equations:

$$C_m \frac{dV_m}{dt} = -\left(E_p + V_m\right) g_p + \left(E_{Na} - V_m\right) g_{Na} - \left(E_K + V_m\right) g_K \qquad (1)$$

Where V_m is the membrane potential, C_m is the membrane capacitance, E_l, E_{Na}, E_K are the Nernst (or reversal) potentials for passive (leaky), Na, and K ion species, and g_p, g_{Na}, g_K are their respective conductances. Whereas this equation was originally derived for a membrane patch, it has been generalized to represent activities of population of neurons:

$$\frac{dx_i}{dt} = -A_i x_i + \left(B_i - x_i\right)\left(\sum_j w_{ji}^{+} f_j\left(x_j\right) + I_i^{+}\right) - \left(D_i + x_i\right)\left(\sum_j w_{ji}^{-} h_j\left(x_j\right) + I_i^{-}\right) \qquad (2)$$

where x_i is the membrane potential of the ith neuron in the network and $f_i(x_i)$ is the output (e.g., in terms of spike frequency) of the ith neuron. A_i is the passive decay constant (originating from the passive (leaky) ion channels), B_i and D_i are the Nernst potentials associated with excitatory and inhibitory channels, respectively and the terms $\left(\sum_j w_{ji}^{+} f_j\left(x_j\right) + I_i^{+}\right)$ and $\left(\sum_j w_{ji}^{-} h_j\left(x_j\right) + I_i^{-}\right)$ represent excitatory and inhibitory conductances, respectively. These conductances are modulated by excitatory signals $f_j(x_j)$ and inhibitory signals $h_j(x_j)$ multiplied by excitatory and inhibitory synaptic weights w_{ji}^{+} and w_{ji}^{-}, respectively. I_i^{+} and I_i^{-} represent external excitatory and inhibitory inputs to the network, respectively. This model is also known as a *shunting* or *multiplicative* model. A simplified version of this model, known as *leaky integrator* model or *additive* model, uses passive rather than active conductances. This model is expressed as:

$$\frac{dx_i}{dt} = -A_i x_i + \sum_j w_{ji} f_j\left(x_j\right) + I_i, \qquad (3)$$

where, whether a signal from cell j to I is excitatory or inhibitory is determined by the sign of the synaptic weight. w_{ji} Similarly, the external input I_i can be positive or negative resulting in excitatory or inhibitory effects, respectively.

For simplicity and for continuity with an earlier masking model proposed by Landahl (1967), Weisstein used the Rashevsky-Landahl neural network mathematical formalism (Weisstein, 1968; Weisstein et al., 1975). This mathematical model, in turn, was inspired by Blair's "theory of excitation" (1932):

$$\frac{dc}{dt} = AI - a\left(c - c_o\right), \tag{4}$$

where c is the average concentration of a positively charged ion, I is the current of that ion, c_0 is the resting concentration of the ion, A and a are positive constants. In this model, "excitation occurs" when, $c > c^\star$, where $c^\star$ is a constant concentration level that represents the threshold of activation. Note that this model captures the role of ionic concentrations in setting the membrane potential and the importance of threshold-like nonlinearity; however, as this model dates back to the 1930s, membrane processes are not incorporated into it. Rashevsky (1933, 1960) adopted and generalized this model by introducing inhibitory effects described by the same type of ionic dynamics. Accordingly, we have:

$$\begin{cases} \dfrac{d\epsilon_t}{dt} = AI - a\left(\epsilon_t - \epsilon_0\right) \\ \dfrac{dj_t}{dt} = BI - b\left(j_t - j_0\right) \end{cases}, \tag{5}$$

where ϵ_t and j_t are excitatory and inhibitory factors, respectively, and the rest of the symbols have the same meaning as Blair's equation. Given that this model takes into account both excitatory and inhibitory effects, it is called the "*two-factor model*". In this model excitation occurs when, $\epsilon \geq j$, i.e., when excitation exceeds inhibition. For convenience, introduce the variables $\epsilon \stackrel{\text{def}}{=} \epsilon_t - \epsilon_0$ and $j \stackrel{\text{def}}{=} j_t - j_0$. Moreover, for neurons that receive their inputs from other neurons in the network, I can be replaced by $f(t)$, which represents the output or firing rate of the pre-synaptic neuron. Accordingly, Equation (5) becomes

$$\begin{cases} \dfrac{d\epsilon}{dt} = Af\left(t\right) - a\epsilon \\ \dfrac{dj}{dt} = Bf\left(t\right) - bj \end{cases} \tag{6}$$

The output of the neuron is determined by a "linear-above-threshold" nonlinearity:

$$f\left(t\right) = \begin{cases} \epsilon - j - h \;\; if\, \epsilon \geq j + h \\ 0 \qquad\qquad if\, \epsilon < j + h \end{cases} \tag{7}$$

In other words, the neuron fires when the net depolarization ($\epsilon - j$) exceeds a threshold *(h)*.

As mentioned above, additive equations are simplified versions of shunting equations. On the other hand, the two-factor model appears to have a different mathematical formalism. However, it was shown that the two-factor model can be expressed as a small circuit consisting of one direct path and one inter-neuron (Figure 6.1) obeying additive equations (Öğmen, 1993).

Another popular neural network model proposed by Bridgeman (1971) emerged from the studies of neural responses in the compound eye of *Limulus*. In a series of electrophysiological experiments, Ratliff, Hartline and Miller (1963) determined that there exists lateral inhibition between these visual neurons and the lateral inhibition is feedback, rather than feed-forward. They used the following static (or steady-state) expression to quantify neural activities:

$$r_i = e_i - \sum_j k_{ji}[r_j - r_{ji}]^+ \tag{8}$$

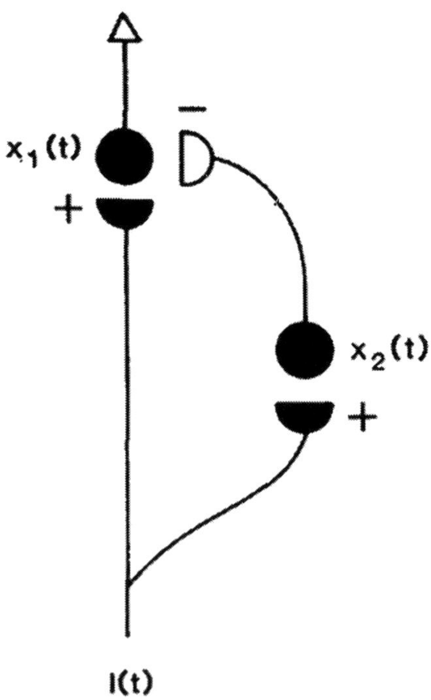

FIGURE 6.1

An additive network consisting of a direct path via neuron x_1 and an inhibitory interneuron x_2 is equivalent to a two-factor neuron.

where e_i and r_i are the external input and the response of the *i*th neuron in the network, r_{ji} is the "firing threshold" for the connection between the *i*th and *j*th neuron, k_{ji} is the synaptic weight for the synapse between the *i*th and *j*th neuron in the network and $[.]^+$ is the linear-above-threshold function, i.e.:

$$[a]^+ \stackrel{\text{def}}{=} \begin{cases} a & \text{if } a>0 \\ 0 & \text{if } a\leq 0 \end{cases} . \tag{9}$$

With k_{ji} chosen as non-negative constants, Equation (8) states that neurons in the network laterally inhibit each other through recurrent connections. The dynamic version of this model was developed by Ratliff, Hartline, and Miller (1963) and was expressed as:

$$r_i(t) = e_i(t) - \sum_j k_{ji} \left[\frac{1}{\tau} \int_0^t e^{-\frac{(t-s)}{\tau}} r_j(s)\, ds - r_{ji} \right]^+ , \tag{10}$$

where τ is a positive constant. This equation is an integral equation, which can be converted to a differential equation through a change of variables to compare it with the basic differential equations used for neural dynamics. Define a new variable as:

$$y_i(t) \stackrel{\text{def}}{=} \frac{1}{\tau} \int_0^t e^{-\frac{(t-s)}{\tau}} r_i(s)\, ds . \tag{11}$$

From this definition, we can determine the time-derivative of this new variable

$$\frac{dy_i}{dt} = -\frac{1}{\tau} y_i + \frac{1}{\tau} \sum_j k_{ji} [y_j - r_{ji}]^+ + \frac{1}{\tau} e_i . \tag{12}$$

Comparing this equation with Equation (3), we can see that this is essentially an additive model. Hence, although the direct mathematical expressions of basic neural-dynamics equations used by Weisstein and Bridgeman look quite different, with some simple mathematical manipulations, it can be shown that they are equivalent. Hence what differentiates between these models is not the fundamental form of the mathematical equations, but rather their architectural and functional forms as well as their "linking hypotheses", i.e., the hypothesis linking the activity in the network to the perceptual and behavioral variables used in psychophysical experiments. In the next section we will discuss these differences.

Architectural and Functional Comparisons of Bridgeman and Weisstein Models

Figures 6.2 and 6.3 show the architectures of Bridgeman's and Weisstein's models, respectively. Bridgeman's model uses a *single channel* in which neurons are interconnected through inhibitory feedback connections (lateral feedback inhibition). When a stimulus is presented to this network, information about the stimulus propagates laterally within the network and due to feedback connections typically generates a standing wave of oscillations. If the stimulus is brief, as in the case of most masking studies, oscillations dampen and dissipate after stimulus offset. In a neural network model for perceptual effects, one needs to formulate a "linking hypothesis", which states how neural activities give rise to percepts. As a linking hypothesis, Bridgeman took the activity generated by the target when presented alone to represent the target's visibility. He assumed that the visibility of the target when presented along with a mask could be based on the similarity between the activity generated by the target and mask in comparison with the activity generated by the target alone. In a first implementation, he used the Pearson r coefficient to quantify the similarity between target-only and target-and-mask induced activities. For reasons that will be discussed in the next section, he later switched to r^2 instead of r.

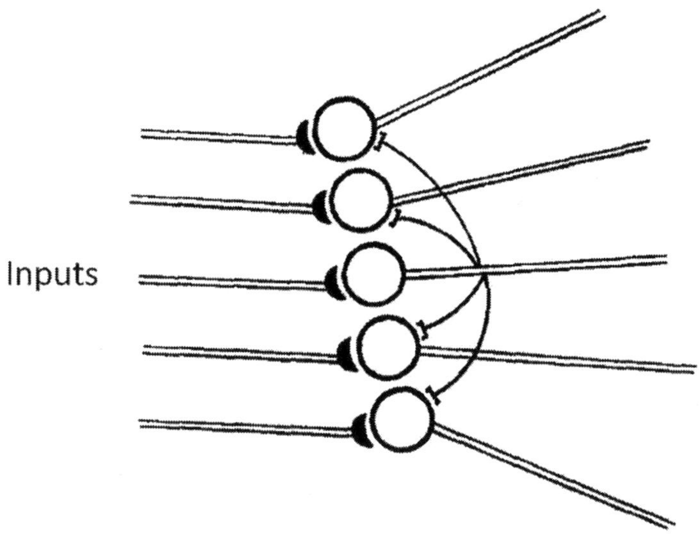

FIGURE 6.2 The architecture of Bridgeman's model.

Input signals are delivered to each neuron through excitatory synapses indicated in black. Every neuron laterally inhibits its neighbors through feedback connections. To avoid clutter, only one set of inhibitory connections are shown in the figure. [From Bridgeman (1971).]

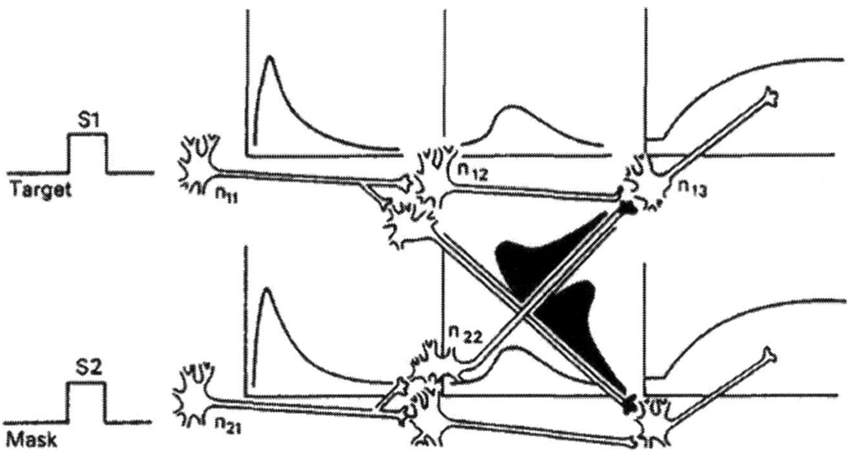

FIGURE 6.3 The architecture of Weisstein's model.

The activities generated by the target and mask at neurons n_{11} and n_{21} excite each two neurons. Excitatory and inhibitory synapses are depicted by open and filled synaptic symbols, respectively. Neuron n_{12} carries target related activities to neuron n_{13} whereas the mask activated neuron n_{22} sends an inhibitory signal to neuron n_{13}. The plots next to each neuron represent the activity of that neuron as a function of time. One can see that activity of neuron n_{22} has shorter latency (i.e., faster) and is more transient than the activity of neuron n_{12}. [From Weisstein et al. (1975).]

Figure 6.3 shows Weisstein's model. The neural network consists of eight neurons. Neurons n_{11} and n_{21} receive the target and mask inputs, respectively. As seen in the figure, the outputs of these neurons bifurcate and give rise to a slow excitatory and a fast inhibitory pathway. Excitatory and inhibitory synapses are indicated by open and filled synaptic symbols, respectively. The inhibitory pathway is faster than the excitatory one, as indicated by the time of occurrence of peak output in each pathway. Hence, in Weisstein's model, we have a feed-forward "*dual channel*" architecture, a slow excitatory channel and a fast inhibitory channel. It is this dual-channel structure that provides a mechanistic explanation for the Type-B masking functions. Since the inhibitory activity coming from the mask-activated pathway to the target-activated pathway, more specifically to neuron n_{13}, is faster than the target-activated input (via neuron n_{12}), the optimal masking is predicted to occur when the presentation of the mask is delayed with respect to the presentation of the target so that the peak of inhibitory activity (depicted in black in Figure 6.3) coincides with the peak of the target-induced activity from neuron n_{12}. While this intuitive explanation provides a simple way of showing how U-shaped functions are produced by this model, subsequent analyses indicated that, for the parameters used in simulations, additional mechanisms, such as mask-blocking, do affect model behavior (Francis, 2000). Fundamentally, the architectures of Bridgeman's and Weisstein's models are different: The former is *recurrent (feedback)* and is *single channel*, whereas the latter is *feed-forward* and *dual channel*.

Dual-channel Models

Another dual-channel model was proposed by Breitmeyer and Ganz (1976), based on psychophysically-determined sustained and transient channels and their putative neural correlates, first the X-Y cell dichotomy of cat retinal cell-types and later a more extended version relying on parvocellular and magnocellular pathways identified in the primate visual system (Breitmeyer & Öğmen, 2006). Although not mathematically formulated, this model provided much more detailed analyses of interactions within and between these dual-channels. Whereas in Weisstein's model the dual channel separation occurs after early common processing stages (neurons n_{11} and n_{21}), the Breitmeyer and Ganz model posits parallel channels emerging already at the retinal level, viz., the sustained retinal ganglion cells tuned to high spatial and low temporal frequencies and transient retinal ganglion cells tuned to low spatial and high temporal frequencies. Furthermore, the Breitmeyer and Ganz model considers inhibitory interactions within (intra-channel inhibition) and between (inter-channel inhibition) channels. In this model, masking can result from intra-channel inhibition, from inter-channel inhibition, as well as by sharing common pathways by the target and mask when they spatially overlap. Figure 6.4 illustrates the Breitmeyer-Ganz dual-channel model.

Another model within this family is the Retino-Cortical Dynamics, or in short the RECOD model (Öğmen, 1993). The starting point of this model was to analyze how cortical networks deal with massive positive feedback. Mathematical and engineering results show that it is extremely difficult to control the dynamics of systems with positive nonlinear feedback. These systems tend to show unstable behavior, meaning that in practice cortical activities may be driven to saturation or into complex oscillatory dynamics. The RECOD model suggests that cortical networks avoid undesirable unstable behavior by operating in a *succession of transient regimes* (Öğmen, 1993; Öğmen et al, 2006a). Accordingly, retino-cortical dynamics are posited to operate in three phases: When a new input is received neurons in the sustained pathway overshoot first and decay to a lower sustained plateau. During the overshoot phase, a strong afferent signal is sent to cortical networks reading-out the information about the new input. This phase is called the *feed-forward dominant phase*. The afferent signal then activates cortical feedback loops. As the afferent signal decays to its plateau, feedback signals start to dominate cortical processing. This second phase of operation is called the *feedback dominant phase*. Finally, when the input changes, a transient signal from the transient pathway inhibits and hence "resets" cortical feedback loops to prevent hysteresis and instability. This phase of operation is called the *reset phase*. The network architecture of the RECOD model is shown in Figure 6.5. Note that, as in the Weisstein and Breitmeyer-Ganz models, it has a dual-channel structure, while adhering more closely to the separate but interacting channel architecture of the Breitmeyer-Ganz model. Unlike the Weisstein and Breitmeyer-Ganz models, this is not a

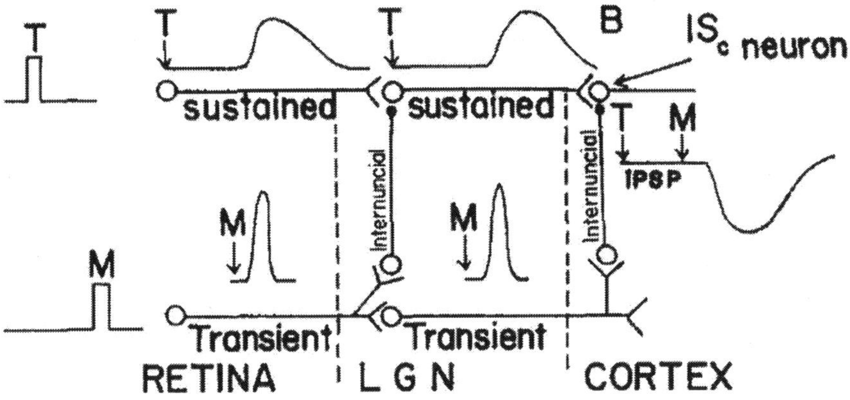

FIGURE 6.4 The architecture of Breitmeyer and Ganz's model.

At the retinal level, target and mask activities are processed by two separate parallel channels consisting of sustained and transient neurons. As depicted in the figure, transient neurons have a shorter latency than sustained neurons and exhibit a more transient response. In this early version of the model, interactions between transient and sustained channels are assumed to take place at the LGN level. In the updated version of the model, these interactions take place at cortical levels. [From Breitmeyer & Granz (1976).]

feed-forward model since it contains feedback (recurrent, reentrant) connections. This aspect of the RECOD model is more similar to Bridgeman's model than the previous dual-channel models. From that perspective, the RECOD model integrates recurrent models of masking with feed-forward models of masking, thereby combining the idea of recurrent interactions for perceptual synthesis with the idea of controlling cortical dynamics through feed-forward interactions between fast and slow channels.

Multiple factors can influence visual masking (Breitmeyer & Öğmen, 2006). It is interesting to highlight here recent findings that show how spatial organization, Gestalt factors, can drastically modify target-mask interactions (Dombrowe et al., 2009; Sayim et al., 2014; cf. Francis & Herzog, 2004). That figural organization can have significant effects on even simple visual tasks, the "*object superiority effect*", was one of Weisstein's pioneering works (e.g., see Chapter 10 (Peterson) in this book). Unfortunately, due to limited computational power, spatial representations in quantitative models of masking have been quite limited. For example, Weisstein's model uses single-point inputs, one for the target and one for the mask; Bridgeman's model and RECOD use *one*-dimensional space. Had Weisstein had the computational resources available today, it would have been a very natural step for her to expand the spatial representations in her model to integrate spatial organization principles.

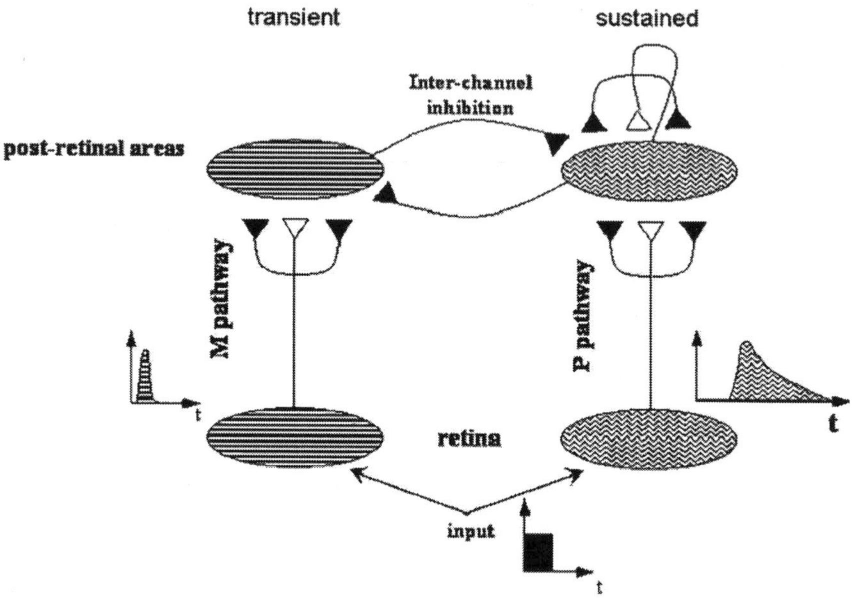

FIGURE 6.5 The architecture of the RECOD model.

The input is processed by two parallel pathways, the magnocellular and parvocellular pathways giving rise to sustained and transient channels. These parallel pathways interact at cortical levels. Filled and open synaptic symbols depict inhibitory and excitatory connections, respectively. Note that the cortical sustained network has feedback connections. [From Öğmen & Breitmeyer (2006), Figure 9.2, p. 154, by permission of The MIT Press.]

Closing Remarks

After its theoretical development, the RECOD model has been used to explain existing masking data as well as to make new predictions for empirical testing. Furthermore, since RECOD is expressed mathematically (as a system of nonlinear differential equations using the aforementioned shunting and additive equations; a full mathematical description of the model can be found in Breitmeyer & Öğmen, 2006), it allowed quantitative simulations that can be directly compared to empirical data. One such work consisted of modeling the phenomenon known as "disinhibition" or "target recovery" in metacontrast. Previous research showed that the visibility of a target (T) masked by a primary mask (M1) can be recovered by the addition of a secondary mask (M2) to the T–M1 sequence (e.g., Briscoe et al., 1983; Byron & Banks, 1980; Dember & Purcell, 1967; Dember et al., 1978; Kristofferson et al., 1979; Long & Gribben, 1969; Purcell & Stewart, 1975; Purcell et al., 1982; Robinson, 1966, 1968; Schiller & Greenfield, 1969; Tenkink & Werner, 1981). Breitmeyer, Rudd, and Dunn (1981) investigated the timing of target recovery in metacontrast using the stimulus shown in Figure 6.6.

They identified two critical properties of target recovery in metacontrast: 1. Maximum target recovery occurs when M2 temporally precedes M1. 2. There is a double dissociation between target recovery and the primary mask's (M1) visibility: While the secondary mask M2 led to the recovery of target's visibility, this happened without a concomitant change in the visibility of the primary mask. Hence, the primary mask M1's masking effectiveness is dissociated from its own visibility. Second, as M2 was delayed further with respect to the target, the opposite dissociative effect was observed: Whereas the visibility of M1 decreased, there was no concomitant change in the visibility of the target. This double dissociation supports the hypothesis that the visibility of the metacontrast mask M1 and its masking effectiveness are carried out by different processes. This outcome is problematic for single-channel models of masking but can be accommodated by dual-channel models. We revisited this phenomenon with two goals in mind: On the one hand, to quantitatively explain target recovery and the aforementioned double dissociation between the mask's visibility and its masking effectiveness and, on the other hand, to test a novel prediction of the RECOD model: Since RECOD hypothesizes that sustained and transient afferent signals emerge from parvocellular and magnocellular pathways, the contrast dependence of metacontrast and target recovery should parallel the contrast dependence of afferent magnocellular and

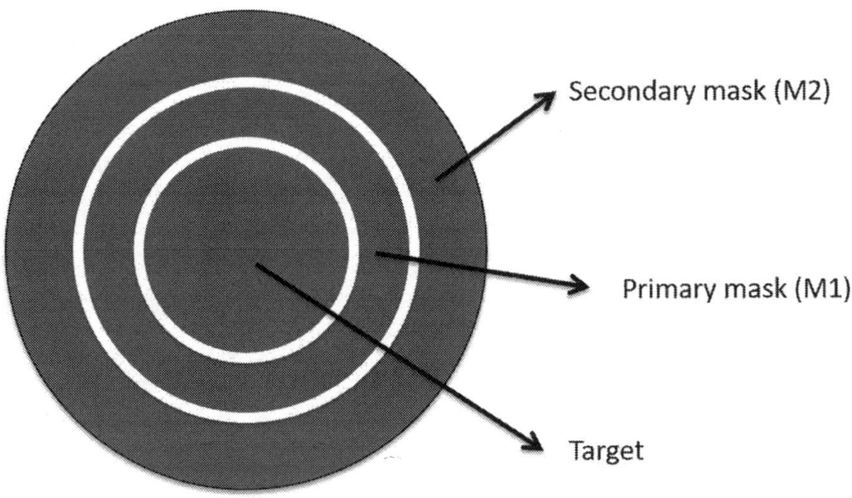

FIGURE 6.6 An example of stimulus used in target recovery experiments.

The central disk is the target stimulus. The ring surrounding this central disk is the primary mask (M1). Typically, the SOA between the target and the primary mask is set to obtain strong suppression of the target's visibility. When a secondary mask (M2) is added to the display, with appropriate timing, it can lead to the recovery of the target's visibility.

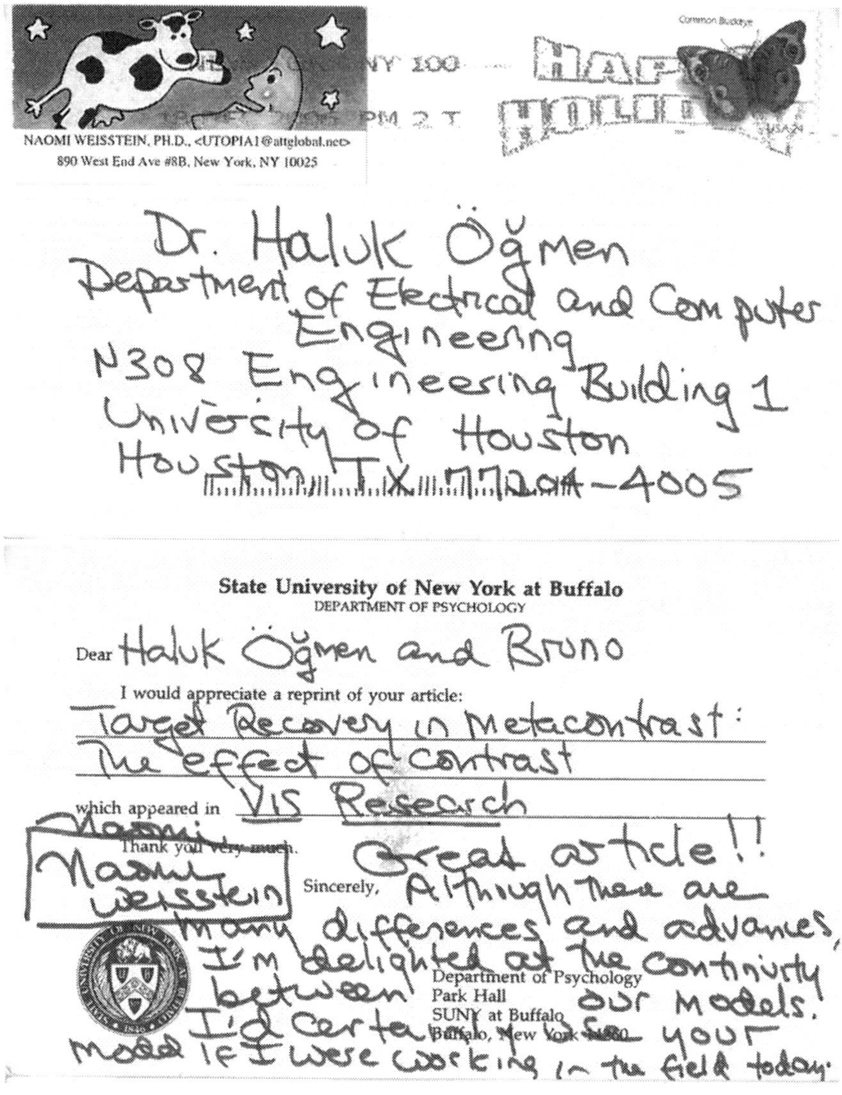

NAOMI WEISSTEIN, PH.D., <UTOPIA1@attglobal.net>
890 West End Ave #8B, New York, NY 10025

Dr. Haluk Öğmen
Department of Electrical and Computer Engineering
N308 Engineering Building 1
University of Houston
Houston, TX 77204-4005

State University of New York at Buffalo
DEPARTMENT OF PSYCHOLOGY

Dear Haluk Öğmen and Bruno

I would appreciate a reprint of your article:
Target Recovery in Metacontrast: The effect of contrast
which appeared in VIS Research

Thank you very much.

Naomi Weisstein

Sincerely,

Great article!! Although there are many differences and advances, I'm delighted at the continuity between our models. I'd certainly use your model if I were working in the field today.

Department of Psychology
Park Hall
SUNY at Buffalo
Buffalo, New York

FIGURE 6.7 Postcard sent by Naomi Weisstein around Christmas 2006.

parvocellular pathways (Kaplan & Shapley, 1986; Purpura et al., 1988), respectively. Empirically, we tested this prediction by systematically varying the secondary mask M2's contrast and the relative timing of the primary and secondary masks (M1–M2 SOA). At the optimal M1–M2 SOA, target recovery increased with M2's contrast without saturating, but at the optimal M1–M2 metacontrast SOA, the reduction of M1's visibility saturated very rapidly as M2's contrast increased. We then carried out simulations of the RECOD model and found a good quantitative agreement with model predictions and experimental data. We published our results in *Vision Research* in 2006 (Öğmen, et al., 2006b). During Christmas holidays in 2006, I found a reprint request in my mailbox (Figure 6.7).

It was a note from Naomi in which she stated "*Great article!! Although there are many differences and advances, I'm delighted at the continuity between our models. I'd certainly use your model if I were working in the field today.*" I remembered reading masking literature during my graduate study days and later I had the opportunity to work closely with Bruno Breitmeyer and met Bruce Bridgeman at various conferences. However, since Naomi was not active due to illness during the times I became active in masking research, I never had the opportunity to meet her in person. Her work was inspiring to me, in particular how she combined advanced computational approaches with experimental work; she was well ahead of her time in both endeavors, in particular in the field later coined as "computational neuroscience". So it was touching and inspiring for me to receive this note from Naomi to realize that I was able to take a journey along the path that she has heroically charted for many generations to follow.

Acknowledgments

I thank Jim Brown and Greg Francis for helpful input.

Note

1. Many readers will recognize a more recent version of this hypothesis, viz., "object substitution masking" (Enns & Di Lollo, 1997).

References

Averbach, E., & Coriell, A. S. (1961). Short-term memory in vision. *Bell System Technical Journal, 40*, 309–328

Bachmann, T. (1994). *Psychophysiology of Visual Masking: The Fine Structure of Conscious Experience*. New York: Nova Science Publishers.

Blair, H.A. (1932). On the measure, of excitability. *Journal of General Physiology, 16*, 165–175.

Breitmeyer, B. G., & Ganz, L. (1976). Implications of sustained and transient channels for theories of visual pattern masking, saccadic suppression, and information processing. *Psychological Review, 83*, 1–36.

Breitmeyer, B. G., & Öğmen, H. (2000). Recent models and findings in visual backward masking: A comparison, review, and update. *Perception & Psychophysics, 62*, 1572–1595.

Breitmeyer, B. G., & Öğmen, H. (2006). *Visual Masking: Time Slices Through Conscious and Unconscious Vision*. Oxford: Oxford University Press.

Breitmeyer, B. G., Rudd, M., & Dunn, K. (1981). Metacontrast investigations of sustained-transient channel inhibitory interactions. *Journal of Experimental Psychology: Human Perception and Performance, 7*, 770–779.

Bridgeman, B. (1971). Metacontrast and lateral inhibition. *Psychological Review, 78*, 528–539.

Briscoe, G., Dember, W. N., & Warm, J. S. (1983). Target recovery in visual backward masking: No clear explanation in sight. *Journal of Experimental Psychology: Human Perception and Performance, 9*, 898–911.

Byron, D., & Banks, W. P. (1980). Patterned stimuli in disinhibition and backward masking. *Bulletin of the Psychonomic Society, 15*, 105–108.

Dember, W. N., & Purcell, D. G. (1967). Recovery of masked visual targets by inhibition of the masking stimulus. *Science, 157*, 1335–1336.

Dember, W. N., Schwartz, M., & Kocak, M. (1978). Substantial recovery of masked visual target and its theoretical interpretation. *Bulletin of the Psychonomic Society, 11*, 285–287.

Dombrowe, I., Hermens, F., Francis, G., Herzog, M. H. (2009). The roles of mask luminance and perceptual grouping in visual backward masking. *Journal of Vision, 9*(11):22, 1–11.

Enns, J. T., & Di Lollo, V. (1997). Object substitution: A new form of visual masking in unattended visual locations. *Psychological Science, 8*, 135–139.

Francis, G. (2000). Quantitative theories of metacontrast masking. *Psychological Review, 107*, 685–768.

Francis, G., & Herzog, M. H. (2004). Testing quantitative models of metacontrast masking. *Psychonomic Bulletin & Review, 11*, 104–112.

Grossberg, S. (1988). Nonlinear neural networks: Principles, mechanisms, and architectures. *Neural Networks, 1*, 17–61.

Kaplan, E., & Shapley, R. M. (1986). The primate retina contains two types of ganglion cells, with high and low contrast sensitivity. *Proceedings of the National Academy of Science USA, 83*, 2755–2757.

Kristofferson, A. B., Galloway, J., & Hanson, R. G. (1979). Complete recovery of masked visual target. *Bulletin of the Psychonomic Society, 13*, 5–6.

Landahl, H. D. (1967). A neural net model for masking phenomena. *Bulletin of Mathematical Biophysics, 29*, 227–232.

Long, N. R., & Gribben, J. A. (1969). The recovery of a visually masked target. *Perception & Psychophysics, 10*, 197–200.

Öğmen, H. (1993). A neural theory of retino-cortical dynamics. *Neural Networks, 6*, 245–273.

Öğmen, H., & Breitmeyer, B. G. (Eds) (2006) *The First Half Second: The Microgenesis and Temporal Dynamics of Unconscious and Conscious Visual Processes*. Cambridge, MA: MIT Press.

Öğmen, H., Breitmeyer, B. G., & Bedell, H. E. (2006a). Dynamics of perceptual epochs probed by dissociation phenomena in masking. In H. Öğmen & B. G. Breitmeyer (Eds), *The First Half Second: The Microgenesis and Temporal Dynamics of Unconscious and Conscious Visual Processes* (pp. 149–169). Cambridge, MA: MIT Press.

Öğmen, H., Breitmeyer, B. G., Todd, S., & Mardon, L. (2006b). Target recovery in metacontrast: The effect of contrast. *Vision Research, 46*, 4726–4734.

Purcell, D. G., & Stewart, A. L. (1975). Visual masking by a patterned stimulus and recovery of observer performance. *Bulletin of the Psychonomic Society, 6*, 457–460.

Purcell, D. G., Stewart, A. L., & Hochberg, E. P. (1982). Recovery and nonmonotone masking effects. *Vision Research, 22*, 1087–1091.

Purpura, K., Kaplan, E., & Shapley, R. M. (1988). Background light and the contrast gain of primate P and M retinal ganglion cells. *Proceedings of the National Academy of Science USA, 85*, 4534–4537.

Rashevsky, N. (1933). Outline of a physico-mathematical theory of excitation and inhibition. *Protoplasma, 20*(1), 42–56.
Rashevsky, N. (1960). *Mathematical Biophysics.* New York: Dover.
Ratliff, F. (1965). *Mach Bands: Quantitative Studies on Neural Networks in the Retina.* San Francisco, CA: Holden-Day.
Ratliff, F., Hartline, H. K., & Miller, W. H. (1963). Spatial and temporal aspects of retinal inhibitory interaction. *Journal of the Optical Society of America, 53*, 110–120.
Robinson, D. N. (1966). Disinhibition of visually masked stimuli. *Science, 154*, 157–158.
Robinson, D. N. (1968). Visual disinhibition with binocular and interocular presentations. *Journal of the Optical Society of America, 58*, 254–257.
Sayim, B., Manassi, M., Herzog, M. H. (2014). How color, regularity, and good Gestalt determine backward masking. *Journal of Vision, 14*(7):8, 1–11.
Schiller, P. H., & Greenfield, A. (1969). Visual masking and the recovery phenomenon. *Perception & Psychophysics, 6*, 182–184.
Tenkink, E., & Werner, J. H. (1981). The intervals at which homogeneous flashes recover masked targets. *Perception & Psychophysics, 30*, 129–132.
Weisstein, N., (1968). A Rashevsky-Landahl neural net: Simulation of metacontrast. *Psychological Review, 75*, 494–521.
Weisstein, N. (1972). Metacontrast. In D. Jameson & L. M. Hurvich (Eds), *Handbook of Sensory Physiology: Visual Psychophysics* (Vol. VII/4, pp. 233–272). New York: Springer-Verlag.
Weisstein, N., Ozog, G., & Szoc, R. (1975). A comparison and elaboration of two models of metacontrast. *Psychological Review, 82*, 325–343.

7

SPATIAL CHARACTERISTICS OF A CONTRAST-COMPARISON PROCESS

S. Sabina Wolfson and Norma Graham
Columbia University

We are interested in the "hidden" intermediate visual processes that are difficult to study because they are far from the retina (light stimulus) and also far from the observer's behavior. There are good reasons to think that there are many intermediate processes and that they are important for perception. These processes were an important part of Naomi Weisstein's interests (see, for example, Weisstein & Harris, 1980; Weisstein et al., 1975).

In the long ages of evolution, eyes occasionally saw (and still occasionally see) a blank, unchanging part of the visual field, for example, a large region of blue sky. But most of the time we are looking at regions of the visual field occupied by spatial texture or forms or patterns. This spatial patterning is almost always changing in time, if only as the result of eye movements. So one might well wonder how the spatial pattern that one has just seen affects the visual processing of the spatial pattern that one sees now. In the course of trying to study something else, we found an effect of this kind – an effect of the preceding pattern's contrast on the current pattern's perceived appearance – that surprised us. This effect of contrast adaptation dramatically increases the visibility of some contrast-defined patterns and dramatically decreases that of others. Initially we half-jokingly referred to this as "Buffy adaptation" as a placeholder because we did not yet understand it. (For the origin of the nickname, see Graham & Wolfson, 2007.) This chapter continues our study of this effect of contrast adaptation.

Figure 7.1 shows a typical trial from the experiments reported here. The observer looks at a fixation screen for a second, then a gray screen for about 500 msec, and then a pattern we will call the *adapt pattern*. (This could also be called a mask pattern as Foley, 2011, does. We briefly discuss the overlapping relationship of the terms "adaptation" and "masking" in the Introduction section of Wolfson & Graham, 2009.) The duration of the adapt pattern is 1 sec. The adapt pattern is followed immediately by a *test pattern*. The duration of the test pattern is approximately 100 msec. The test pattern is followed immediately by a *post-test pattern* that is exactly the same as the adapt pattern. The screen then returns to gray (for at least 100 msec) until the observer makes a response.

The adapt pattern is a 2x2 grid of Gabor patches all of the same contrast. (The contrast of a Gabor patch is expressed as one-half the difference between peak and trough luminances divided by the mean luminance underneath the Gaussian window of the Gabor patch.)

Each test pattern is composed of two different contrasts of Gabor patches (which vary from trial to trial). The observers' responses indicate whether they think the contrast-defined stripes in the test patterns are "horizontal" or "vertical", as illustrated on the right side of Figure 7.1. Feedback is provided as to the correctness of the response on each trial. After an observer responds, the screen remains gray until the observer initiates the next trial with a key press.

The spatial frequency of the sinusoidal fluctuation in the Gabor patches is always 2 cycles/degree in the work we report here. There is approximately one full cycle in each Gabor patch. The mean luminance of the screen stays constant throughout the experiment at about 50 cd/m^2.

In the *baseline condition* in our experiments, the test pattern on any trial is exactly like the adapt pattern (and post-test pattern) except for Gabor-patch contrasts. This is the condition illustrated in Figure 7.1.

Many details of the experiments and patterns are given in Table 7.1. The table lists experimental parameters that were not necessarily the same for each experiment. The following parameters were the same for all experiments in this chapter: grid size = 2x2 Gabor patches; Gabor patch spatial frequency = 2 cycles/degree; viewing distance = 90 cm with unrestrained head; adapt contrast = 50 percent; adapt pattern's duration = 1 sec; test pattern's duration = 94 msec; post-test

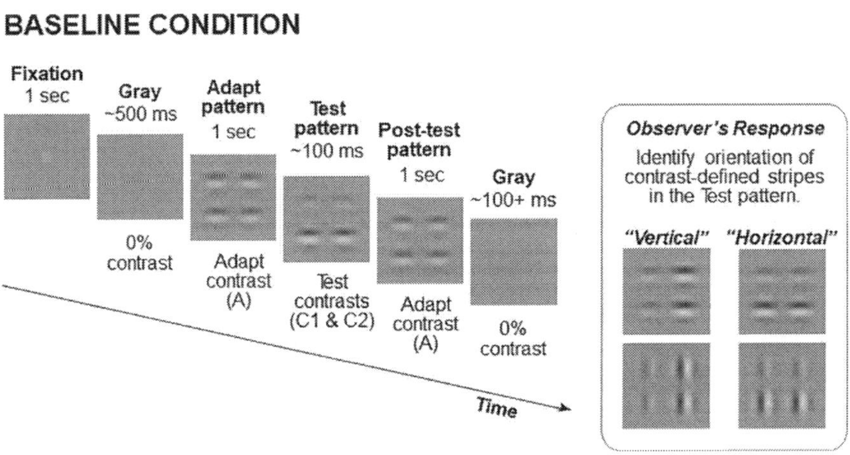

FIGURE 7.1 A typical trial in the *baseline condition*. See text for description.

TABLE 7.1

Plotted in…	*Adapt pattern's Gabor patch orientation (Øadapt)*	*Orientation of test pattern's Gabor patches*	*Phase of test pattern's Gabor patches*	*Position of test pattern's Gabor patches*	*Test contrast differences*	*Average test contrasts |C1-C2|*	*Center-to-center Gabor patch distance*	*Observers*	*Trials per session*	*In house expt name*
ORIENTATION-CHANGE EXPERIMENT - Measuring performance										
Figure 4	0° or 90°	Øadapt +/- 0°, 2°, 5°, 10°, 15°, or 90°	same as adapt pattern's	same as adapt pattern's	10% 20% 30%	one Above, one Straddle, and one Below for each |C1-C2|	2 deg	LG MC	216	BG1orient4i89
ORIENTATION-CHANGE EXPERIMENT - Measuring threshold										
Figure 5	0° or 90°	Øadapt +/- 0° or 90°	same as adapt pattern's	same as adapt pattern's	varied to determine threshold	40% 50% 60%	1 deg	LG BSG LG BSG	90	HGP4i101b H4i101b
PHASE-CHANGE EXPERIMENT - Measuring threshold										
Figure 6	0° or 90°	same as Øadapt	same as adapt pattern's or flipped (180° phase change)	same as adapt pattern's	varied to determine threshold	40% 50% 60%	1 deg	MC BSG MC BSG	90	HG1phase4i101b H4i101b
HORIZONTAL-POSITION CHANGE EXPERIMENT - Measuring performance										
Figure 7	0° or 90°	same as Øadapt	same as adapt pattern's	shift to left or right by 0/32, 2/32, 5/32, 8/32, 16/32, 19/32, 24/32, 29/32, or 32/32 cycles	10% 20% 30%	one Above, one Straddle, and one Below for each |C1-C2|	1 deg	LG MC	306	BJ4i85

pattern's duration = 1 sec. Within each experiment, trials from the baseline condition (in which the adapt, test, and post-test patterns are identical except for contrast) and the other conditions were randomly intermixed. In the horizontal-position-change experiment we used all of the Δposition values listed but we did not plot all of them in Figure 7.7. The rightmost column lists the experiment names we used in the lab in order to allow cross-publication comparison when the same data gets used in different analyses.

In the *baseline condition* in our experiments, the test pattern on any trial is exactly like the adapt pattern (and post-test pattern) except for Gabor-patch contrasts. This is the condition illustrated in Figure 7.1.

Three example test patterns are shown in Figure 7.2. The two test contrasts (C1, C2) always differ by 10 percent, but the average test contrast varies. The test pattern in the center has an average test contrast of 50 percent. In this test pattern, the two test contrasts were 45 percent and 55 percent. These test contrasts *Straddle* the adapt contrast of 50 percent. The test pattern on the left has an average test contrast of 45 percent; the two test contrasts are 40 percent and 50 percent. This average test contrast is *Below* the adapt contrast. The test pattern on the right has an average test contrast of 55 percent; the two test contrasts are 50 percent and 60 percent. This average test contrast is *Above* the adapt contrast.

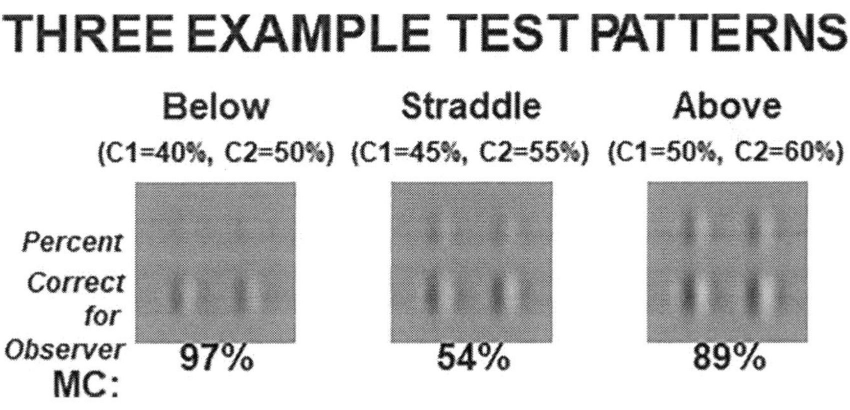

FIGURE 7.2 Three example test patterns, each with a 10 percent test-contrast difference (difference between test contrast C1 and test contrast C2). Grayscale differences have been exaggerated for visual clarity.

The average test contrast of the Below test pattern (45 percent) is "below" the adapt contrast of 50 percent. The average test contrast of the Above test pattern (55 percent) is "above" the adapt contrast of 50 percent. In the Straddle test pattern, the two test contrasts (45 percent and 55 percent) "straddle" the adapt contrast of 50 percent. Performance of observer MC on these test patterns is shown below the grayscale images. While all of these test patterns have the same test-contrast difference, performance on the Straddle test pattern is much worse than performance on the Below or Above test patterns. (The performances here are shown as open circle symbols in the upper left panel of Figure 7.4.)

Result from one observer (MC) in a baseline condition are shown below the test patterns in Figure 7.2 as percent correct. Performance on the Straddle test pattern was worse than for the Below and Above test patterns. We call this effect the *Straddle Effect* and have suggested a contrast-comparison process described later in this chapter to explain this effect.

Foley (2011), using stimuli quite like ours (composed of Gabor patches), also found the Straddle Effect and expanded on our understanding of it. Kachinksy, Smith, and Pokorny (2003), using quite different stimuli (composed of solid squares), and with a different goal (investigating magnocellular vs parvocellular pathways), included contrast conditions similar to ours; if their Figure 6 data is plotted like ours, it shows the Straddle Effect.

Purpose of this Chapter

The purpose of the present chapter is to explore the spatial characteristics of the Straddle Effect in order to elaborate on its hypothesized explanation, the contrast-comparison process.

In all of the experiments reported here, the adapt contrast is always 50 percent and only three kinds of test patterns are used: Below, Straddle, and Above (analogous to those in Figure 7.2). Previously, we have studied the baseline condition with other adapt contrasts and fuller ranges of test contrasts and we have always found the Straddle Effect (see, for example, Wolfson & Graham, 2009).

We will describe several new experiments here, each of which compares the baseline condition to conditions in which the adapt and test patterns differ in their spatial characteristics. The orientation of the test pattern's Gabor patches can differ from the orientation of the adapt pattern's Gabor patches (top panel Figure 7.3); or the phase of the adapt and test patterns' Gabor patches can differ (middle panel Figure 7.3); or the horizontal position of the adapt and test patterns' Gabor patches can differ (bottom panel Figure 7.3).

Results from the Orientation-Change Experiments

Figure 7.4 shows results from two observers (MC and LG) for an orientation change (Δorient) of 0 degrees (the baseline condition) in the leftmost column through an orientation change of 90 degrees in the rightmost column. (An orientation change of 90 degrees is illustrated in the top panel of Figure 7.3.) The open circles (gray squares, black diamonds) show results when the test-contrast difference was 10 percent (20 percent, 30 percent). In each panel, the adapt contrast of 50 percent is marked by an asterisk on the horizontal axis. The points plotted above the asterisk are results from Straddle test patterns. Points to the left are from Below test patterns; points to the right are from Above test patterns.

For each curve in Figure 7.4, the Straddle Effect is clear in the baseline condition: performance on the Straddle test pattern is lower than performance on the

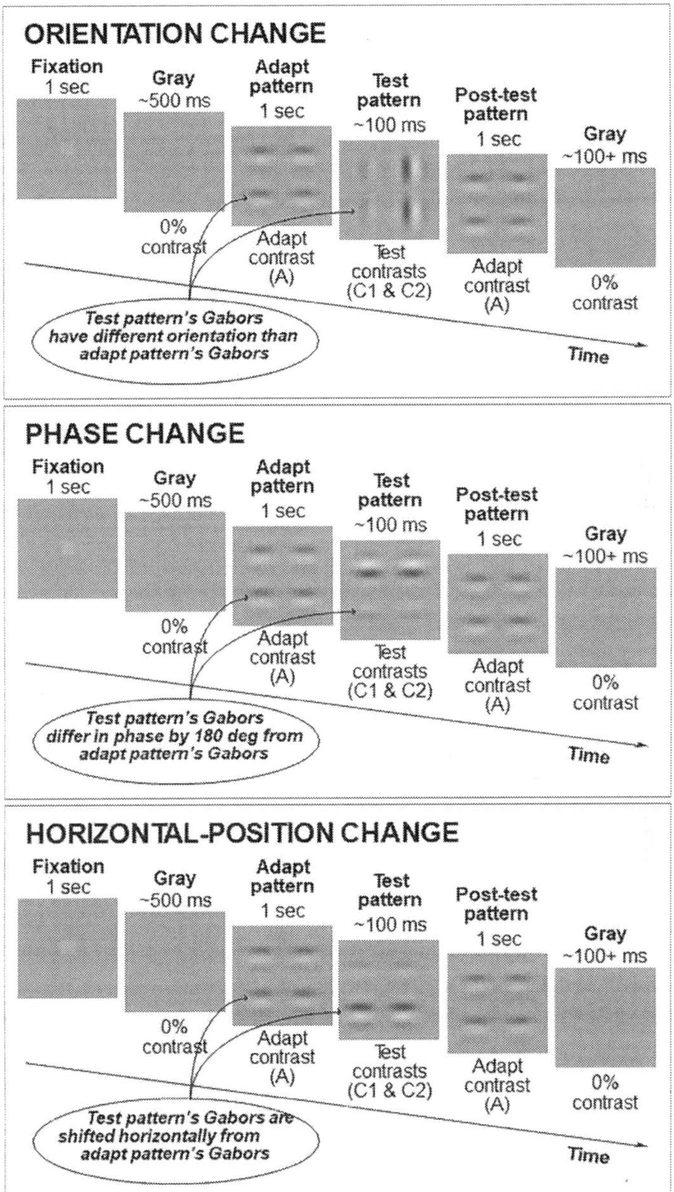

FIGURE 7.3 A typical trial for each of three kinds of experiments: Orientation change (top panel), phase change (middle panel), and horizontal-position change (bottom panel).

These are like the baseline condition except that the test pattern's Gabor patches differ in spatial characteristics from the adapt pattern's Gabor patches. The observer's response – as in the baseline condition – indicates whether he/she thinks the contrast-defined stripes in the test pattern are "horizontal" or "vertical". Feedback (as to the correctness of the response) was provided.

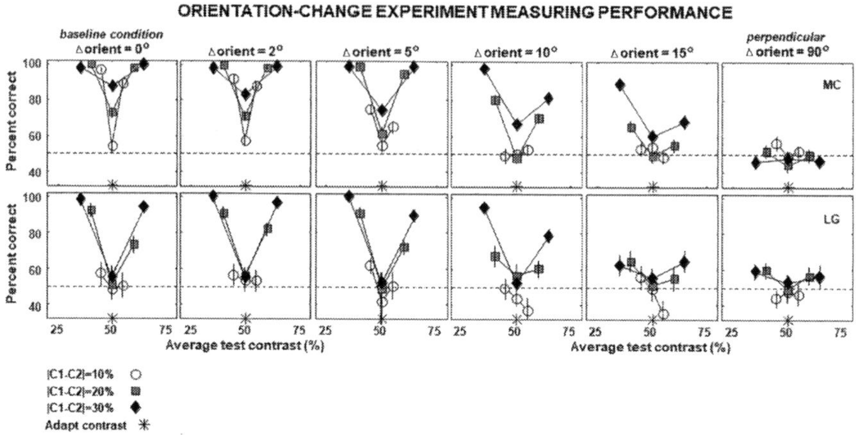

FIGURE 7.4 Results from the orientation-change experiment measuring performance.

The orientation of the Gabor patches in the adapt pattern and test pattern differ by Δorient (which is shown above each column). Average test contrast is shown on the horizontal axis. The asterisk marks the adapt contrast (50 percent) on the horizontal axis. Performance, measured as percent correct, is shown on the vertical axis. Two observers (MC and LG) are shown in the two rows. Error bars show +/− 1 SE across sessions.

The open circle symbols in the upper left panel repeat the results shown in Figure 7.2.

The precise test contrast values we used are as follows: in each panel, the three points with a test-contrast difference (|C1-C2|) of 10 percent (open circle symbols) had average test contrasts of 45 percent, 50 percent, and 55 percent (reading left to right); the points with |C1-C2| = 20 percent (gray square symbols) had average test contrasts of 40 percent, 50 percent, and 60 percent; and the points with |C1-C2| = 30 percent (black diamonds) had average test contrasts of 35 percent, 50 percent, and 65 percent. For example, the test pattern with an average test contrast of 45 percent on the curve for a 10 percent test-contrast difference had C1=40 percent and C2=50 percent.

Below and Above test patterns. (The open circle symbols in the upper left panel correspond to the performance values listed in Figure 7.2.) For each observer, the Straddle Effect becomes less dramatic as the orientation change becomes larger. This is *not* because performance on the Straddle test pattern becomes better but because performance on the Below and Above test patterns becomes worse.

In a second orientation-change experiment, we measured thresholds (Figure 7.5). For each threshold, we held constant the average test contrast and varied the two test contrasts (C1 and C2) symmetrically until performance reached 80 percent correct. The average test contrast was held constant at 40 percent, 50 percent, or 60 percent for the Below, Straddle, or Above test patterns, respectively. Figure 7.5 shows thresholds for the baseline condition (0 degree orientation change, left column) and for the perpendicular condition (90 degree orientation change, right column). Again, the Straddle Effect is present in the baseline condition: the threshold for the Straddle test pattern is higher than those for the Below and Above test patterns. In the perpendicular condition, the Straddle Effect has

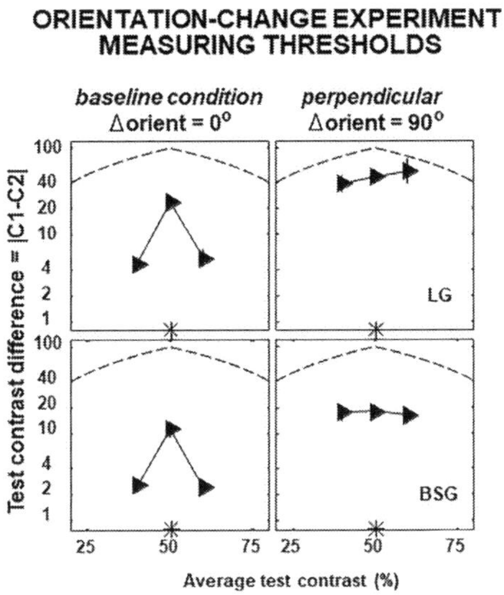

FIGURE 7.5 Results from the orientation-change experiment measuring threshold (rather than performance which is shown in Figure 7.4).

The left column shows results from the baseline condition (orientation change of 0 degrees) and the right column shows results from an orientation change of 90 degrees. Two observers (LG and BSG) are shown in the two rows. The vertical axis shows threshold as the test-contrast difference (|C1-C2|) that leads to 80 percent correct. The horizontal axis shows average test contrast; the three average tests contrasts used were 40 percent, 50 percent, and 60 percent. The asterisk marks the adapt contrast (50 percent) on the horizontal axis. The data points and error bars show the mean +/– 2 SE across sessions. The dashed line shows the maximum possible test-contrast difference.

disappeared because the thresholds on the Below and Above test patterns have become higher.

A Note to Prevent Confusion

Foley (2011) reported an orientation experiment (his Figure 10) which looks superficially like ours, but is in fact a very different experiment. Importantly, he did not keep average test contrast constant while using Below, Straddle, and Above test patterns. Also, his adapt contrast (16 percent) was much lower than ours (50 percent).

Results from the phase-change experiment

In the phase-change experiment, we compared the baseline condition with one phase-changed condition. This phase change – which we call "flipped phase" – is illustrated in Figure 7.3, middle panel; the black and white bars of the Gabor patch switch position between the adapt pattern and test pattern. Figure 7.6, formatted

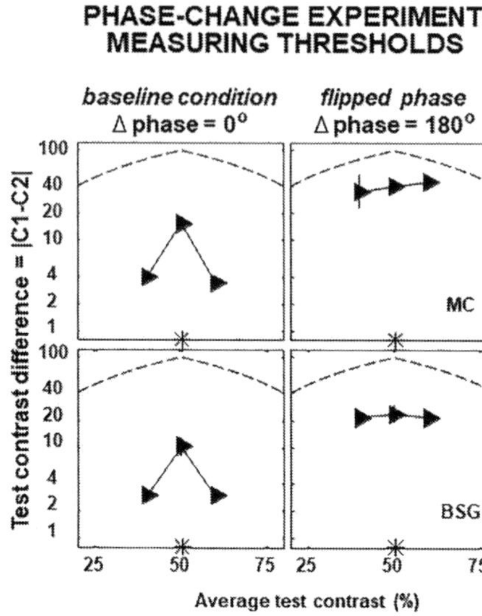

FIGURE 7.6 Results from the phase-change experiment measuring threshold.

The left column is the baseline condition and the right column is the flipped phase condition (in which the phase has been changed by 180 degrees). This figure is in the same format as Figure 7.5.

like Figure 7.5, shows results for the baseline condition (i.e., a phase change of 0 degrees, left column) and for the flipped-phase condition (a phase change of 180 degrees, right column). Again we see the Straddle effect in the baseline condition, but we do not see the Straddle effect in the flipped-phase condition. Again this is because thresholds for the Below and Above test patterns increase to approximately equal that for the Straddle pattern.

Results from the Horizontal-Position-Change Experiment

A horizontal-position change is shown in Figure 7.3, bottom panel; the Gabor patches in the illustrated test pattern are moved left relative to those in the adapt pattern. Figure 7.7, formatted like Figure 7.4, shows results in the baseline condition (leftmost column) and for many different horizontal-position changes (Δposition) up to a maximum of 1 cycle (rightmost column). For each observer, the Straddle Effect becomes less dramatic as the horizontal-position change becomes larger. Again, this is *not* because performance on the Straddle test pattern becomes better but because performance on the Below and Above test patterns becomes worse.

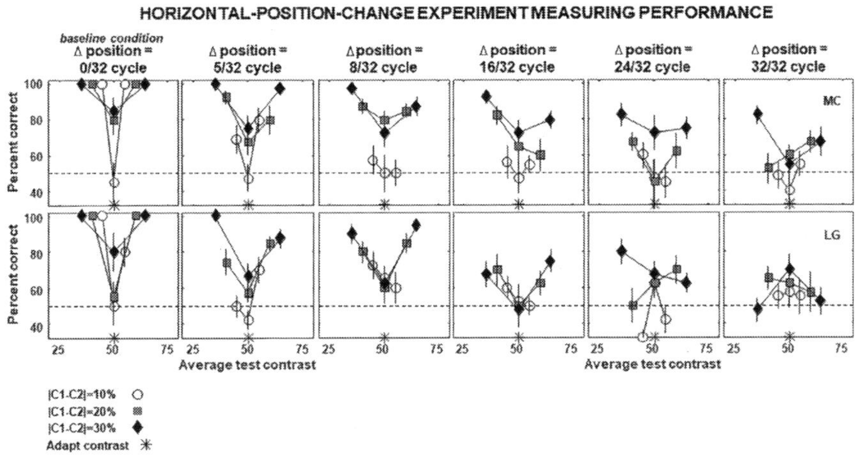

FIGURE 7.7 Results from the horizontal-position-change experiment measuring performance.

The horizontal position of the Gabor patches in the adapt pattern and test pattern differed by Δposition (which is shown above each column). Δposition is measured as fractions of a cycle in the Gabor patches' sinusoidal fluctuation. This figure is in the same format as Figure 7.4.

Note About One Complexity in the Horizontal-Position-Change Experiment

In the baseline condition, the Gabor patches were vertical on half of the trials and horizontal on the other half of the trials. (This was also true in the orientation-change experiments and the phase-change experiment but was not a confounding factor in those experiments.) It might be a confounding factor in the horizontal-position-change experiment because the position change was always horizontal, never vertical. However, as it turns out, results for the vertical Gabor patch trials considered alone look very similar to results for the horizontal Gabor patch trials considered alone. Thus, we have not separated the trials in Figure 7.

Explanation for the Straddle Effect that *Does Not* Work

Consider again the test patterns show in Figure 7.2. The test-contrast difference is 10 percent for each test pattern. However, the *change* in a single Gabor patch's contrast at the transition between the adapt pattern and test pattern is +5 percent or –5 percent for the Straddle test pattern, while the change is 0 percent or –10 percent for the Below test pattern (0 percent or +10 percent for the Above test pattern). Thus one might think that the poorer performance on the Straddle test pattern is because the individual changes are of smaller magnitude (5 percent) than the maximum change on the Below and Above test patterns (10 percent).

To test this idea, we need to consider performance on a Straddle test pattern in which the individual changes are of magnitude 10 percent. The data shown in Figure 7.2 are the open circles from the upper left panel in Figure 7.4, so we have such a test pattern: a Straddle test pattern composed of 40 percent and 60 percent contrast Gabor patches (middle gray square plotted above the asterisk). On this Straddle test pattern, observer MC performs at 73 percent correct. This performance (73 percent) is worse than his performance on the Below and Above test patterns (97 percent and 89 percent) with the same maximum magnitude of change (open circle symbols to the left and right of the asterisk). Unfortunately, because LG's performance level shown by the open circle symbols is so low, we cannot see an example of this in her results in this experiment, but we have shown this for many observers before (Wolfson & Graham, 2009, Figure 7).

Another Explanation for the Straddle Effect that *Does Not* Work

One of the reviewers suggested an alternate explanation that is totally reasonable given the results shown in this paper. Here, only three test patterns (Below, Straddle, and Above) are used. Further, in all these Below and Above test patterns only 2 of the 4 Gabor patches have a different contrast than the adapt contrast; the other 2 Gabor patches have the same contrast as the adapt contrast. The reviewer therefore suggests that perhaps it is the *number* of changing contrast levels, rather than their *signs*, that causes the Straddle Effect. The reviewer suggests that a suitable control would be having the contrast of all 4 Gabor patches in the Above test pattern be above the adapt contrast (similarly, all 4 in the Below test pattern be below the adapt contrast). We have frequently collected such data (e.g. Figure 12 and Figure A4 of Wolfson & Graham, 2009) and these data still show the Straddle Effect (as long as the test contrasts are not all very far below or all very far above the adapt contrast).

Explanation for the Straddle Effect that *Does* Work

The Straddle Effect can be explained by a process (which we call contrast comparison) in which the sign of a contrast change is lost (or very degraded), but the magnitude of that contrast change is preserved. It is as if

- Adaptation to contrast resets some comparison level to equal the recent time-averaged contrast seen at that location in the visual field.
- The current contrast at each spatial position is evaluated relative to its comparison level, with increases and decreases being equally effective but increases being quite confusable with decreases (within whatever visual subsystem is responsible for the observer's performance in our experiments).

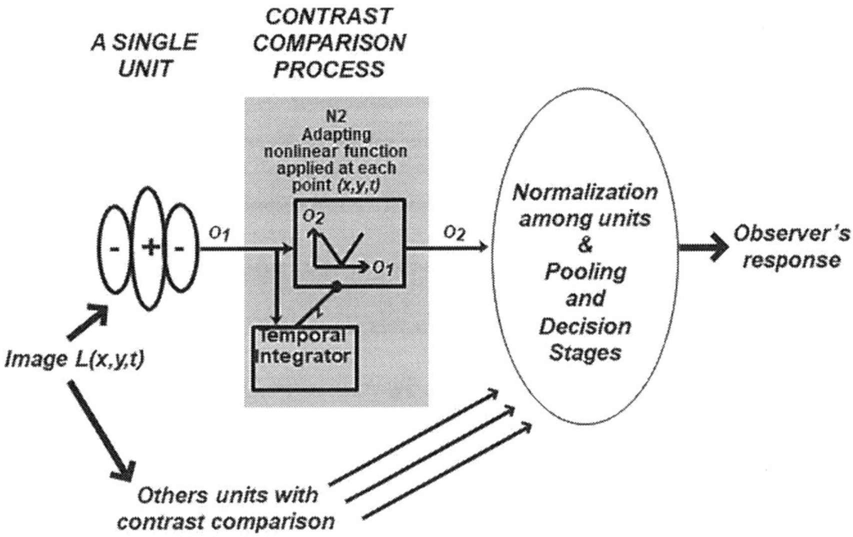

FIGURE 7.8 Our overall model framework showing the shifting, rectifying contrast-comparison process and the normalization process.

The normalization process will not be discussed here; it is a known, important, "hidden" intermediate process (see, e.g., Reynolds & Heeger, 2009, and Solomon & Kohn, 2014). The Pooling and Decision Stages need to include a process or processes that allow the observer to tell the difference between horizontal and vertical arrangements of contrast. This figure adds a typical V1 receptive field as input to the contrast-comparison process within the framework we have shown before (e.g. Figure 17 in Graham, 2011). The addition of a typical V1 receptive field is consistent with the results of this chapter.

Such a contrast-comparison process is diagramed in the gray box in Figure 7.8.

In other words, there exists something in the visual system that acts like a rectification function on the contrast dimension making it difficult to perceive a Straddle test pattern correctly. Consider the extreme case of a full-wave rectification as shown in Figure 7.8. In that case, when a Straddle test pattern appears, the Gabor patches which increase in contrast produce the same output as the Gabor patches which decrease in contrast. The observer *can* see that each Gabor patch has changed, but the observer *cannot* identify the sign of the change in each Gabor patch. Thus, it looks like all four Gabor patches change but all the changes look the same. So the observer cannot see whether the second-order stripes are horizontal or vertical. A full-wave rectification is more extreme than we have found when modeling the results of any observer's performance, but less extreme rectifications have always worked (see Graham & Wolfson, 2013).

For a systematic and careful study of the subjective phenomena in experiments like ours, see Foley (2011).

Helpful or Hurtful?

While it is easy to think of the Straddle Effect as oddly poor performance on Straddle test patterns, we suspect it is more accurately described as a side effect of a generally advantageous process. And we suspect that this generally advantageous process is one that enables humans to detect changes well.

We've explained the Straddle Effect by assuming that the magnitude of the change is encoded well but that the sign of the change is not. Why might the sign be encoded poorly? A change that produces a ringing output is easy to detect in any number of ways, but it is hard to know the direction (i.e., the sign) of change because that requires knowing whether the first excursion was up or down. For other reasons why the sign might be encoded poorly, see Graham and Wolfson (2013).

In Closing

We found that when the spatial characteristics of the adapt and test patterns were identical, there was always a Straddle Effect. But when the adapt and test patterns differed in orientation, phase, or position, the Straddle Effect was much diminished. This diminishing was *not* because performance on the Straddle test patterns improved, but because performance on the Below and Above test patterns got worse.

We currently think that the Straddle Effect is likely to be a side effect of the evolutionary importance of detecting changes/transients.

We initially ran across the Straddle Effect by accident while exploring the dynamics of contrast adaptation using the method that we had previously used to explore the dynamics of light adaptation (Graham & Wolfson, 2007). In those initial experiments we used a large, 15x15 grid of Gabor patch elements, thinking that the Straddle Effect was a global second-order process. As we explored the effect more, we began to wonder if such a large grid was necessary. We tried a 2x2 grid and still found the Straddle Effect, and thus we knew it was more local than we had initially thought. The results reported here suggest that the process producing the Straddle Effect could be as local as a typical V1 receptive field shown in Figure 7.8.

Acknowledgements

This work was supported in part by National Eye Institute Grant EY08459. We thank Ian Kwok and Megan Felder, both of whom did some modeling which is not discussed explicitly here but which has been very important for the development of the ideas. We also thank our observers for their hours of effort.

Norma would like to express her gratitude to Naomi not only for her published work but for hours of conversation, and Sabina wishes she had had an opportunity to know her.

References

Foley, J. M. (2011). Forward-backward masking of contrast patterns: The role of transients. *Journal of Vision, 11*(9), 15.

Graham, N. (2011). Beyond multiple pattern analyzers modeled as linear filters (as classical V1 simple cells): Useful additions of the last 25 years. *Vision Research, 51*, 1397–1430.

Graham, N., & Wolfson, S. S. (2007). Exploring contrast-controlled adaptation processes in human vision (with help from Buffy the Vampire Slayer). In L. R. M. Harris & M. R. Jenkin (Eds). *Computational Vision in Neural and Machine Systems* (pp. 9–47). Cambridge: Cambridge University Press.

Graham, N., & Wolfson, S. S. (2013). Two visual contrast processes: One new, one old. In C. Chubb, B. A. Dosher, Z.-L. Lu, & R. M. Shiffrin (Eds). *Human Information Processing: Vision, Memory, and Attention*. Washington, DC: American Psychological Association.

Kachinsky, E. S., Smith, V. C., & Pokorny, J. (2003). Discrimination and identification of luminance contrast stimuli. *Journal of Vision, 3*(10), 599–609.

Reynolds, J. H., & Heeger, D. J. (2009). The normalization model of attention. *Neuron, 61*, 168–185.

Solomon, S. G., & Kohn, A. (2014). Moving sensory adaptation beyond suppressive effects in single neurons. *Current Biology, 24*, R1012–R1022.

Weisstein, N., & Harris, C. S. (1980). Masking and the unmasking of distributed representations in the visual system. In C. S. Harris (Ed.) *Visual Coding and Adaptability* (pp. 317–364). Hillsdale, NJ: Lawrence Erlbaum Associates.

Weisstein, N., Ozog, G., & Szoc, R. (1975). A comparison and elaboration of two models of metacontrast. *Psychological Review, 82*(5), 325–343.

Wolfson, S. S., & Graham, N. (2007). An unusual kind of contrast adaptation: Shifting a contrast comparison level. *Journal of Vision, 7*(8), 1–7.

Wolfson, S. S., & Graham, N. (2009). Two contrast adaptation processes: Contrast normalization and shifting, rectifying contrast comparison. *Journal of Vision, 9*(4), 1–23.

8

HOW TO FIND A YELLOW VOLKSWAGEN

Jeremy Wolfe
Harvard Medical School

The Women's Movement came to my corner of Bell Labs in the form of Naomi Weisstein in the early 1970s. My father had arranged a summer job for me with John Krauskopf, a notable color vision researcher and tennis buddy of my father's. Working at Bell changed the trajectory of my life, turning me into an aspiring vision researcher. The place was packed with important figures in vision and cognition; Bela Julesz, Saul Sternberg, Charlie Harris, et al. In my memory, all were males and, in my memory, all had a female research assistant often referred to as "the girl" as in "My girl will bring you the data". The arrival of Naomi Weisstein on what I think was a sabbatical visit stirred things up. I don't recall that anyone had been actively opposed to women's rights. It just hadn't really had much impact yet. Naomi was the personification of "impact". I imagine that the senior scientists felt that impact in a variety of settings. I mostly experienced it at lunch where casually sexist language and assumptions were, shall we say, not going to be ignored. For a 17-year-old kid from the New Jersey suburbs, it was eye-opening.

My time at Bell opened my scientific eyes as well as my political eyes. Before I made it to college, I was introduced to most of the issues that have occupied my scientific career. In that period of my early education, Weisstein was not only my introduction to the issue of the role of women in scientific life. She was also my introduction to object perception, the topic I will discuss in this chapter. She and Charlie Harris were working on the object superiority effect (Weisstein & Harris, 1974). She was also interested in the problem of object recognition. In the early 70s it was discovered that cells in the inferotemporal cortex of the macaque brain, a region known to be devoted to visual processing, often responded best to very specific stimuli like faces or even toilet brushes (Gross et al, 1972). This led to the idea that recognition might result from lighting up one of these specific cells (Barlow, 1995). Barlow talked about "canonical cells" that would represent specific objects. A catchier term, "grandmother detector" was probably coined by Jerry Lettvin (1995) in a letter to Barlow. Weisstein's version was a "yellow Volkswagen" detector. Yellow Volkswagen "Beetles" were popular but not critical to the issue at hand (see Figure 8.1[1]). Her term did make it clear that these detectors could not be a fixed, innate set since it was unlikely that evolution would have encoded automobile models into the genetic code for the visual system.

One of the problems with the idea that activity in a specific neuron (or even a group of neurons) represented the recognition of an object is that we are good at recognizing yellow Volkswagens in an essentially infinite array of sizes, positions, orientations, and so forth. How could you create such a generalized "template" for an object out of the outputs of cells, like those in the primary visual cortex, that respond to oriented lines and edges in very specific parts of the visual field. Interestingly, the rise of deep learning networks (DNNs), developed by computer science experts, has provided a type of existence proof that something like this is possible. Modern object recognition DNNs learn to classify millions of images into, say, 1000 categories (poodle, yellow Volkswagen, etc.). They get "trained" by exposure to a large number of images—analogous, perhaps, to the training of our visual systems as we are exposed to the world. Once trained, a DNN often has early layers that look quite a lot like early stages in the visual system with nodes responding to small localized features and a final set of 1000 recognition nodes that could be thought of as yellow Volkswagen detectors (Krizhevsky et al., 2012). A characteristic of these final nodes is that they will respond to their preferred object in a wide range of poses and settings, just as a yellow Volkswagen detector should. Importantly, the network – like a human – will respond to views and instances of objects that it has never seen before.

FIGURE 8.1 A 1973 yellow Volkswagen © Chris Vallone, Classic VW BuGs.

My particular interest in this topic, and the subject of this chapter, concerns how observers can make use of object recognition processes to find a target object or objects in a visual search task. How do you find a yellow Volkswagen or any other object that you might be looking for in a world full of distractor objects that you are not looking for? There are objects out in the world. Do they match the identity of target objects, held in our mind? In the search literature, we tend to call this mental representation the "search template". The term, "template" conjures up something like a stencil that is an exact match to its item. In some simple tasks, the search template could almost be like such a template. If you are looking for a red vertical line of some specific size, one could imagine a search template that was some sort of exact match to this target (albeit somehow embodied in neurons). More generally, the search template must be more like the grandmother or yellow Volkswagen detector. If you were asked to search for that vehicle or that grandmother, candidate objects in the world would need to be compared to a representation in the mind that could recognize the target over a wide range of variations. What do we know about how that process works? In this chapter, I will briefly discuss five problems:

1) How rapidly can you search through objects?
2) How can you search through objects so rapidly?
3) Even at that speed, how can you find anything in a real scene given how many objects might be in that scene? (And what is an "object" anyway?)
4) What happens if you are looking for more than one type of target at one time?
5) Finally, where in the mind might that yellow Volkswagen detector reside? A leading proposal is that the search template resides in Working Memory (Olivers et al., 2011). This chapter will suggest an alternative account.

Problem One

How rapidly can you search through objects? In Figure 8.2, look for a beetle. It won't be hard to find even though we did not specify the exact beetle. You can select items from this display, compare them to some sort of beetle template in your mind and answer yes or no (or, perhaps, "maybe" if you stumbled upon that scorpion).

In a standard visual search paradigm (Treisman & Gelade, 1980) (Wolfe et al., 2010), observers (Os) would be shown a search stimulus and asked to respond with a keypress as rapidly and accurately as possible as to whether the designated target was or was not visible. Typically, the target would be present on 50 percent of the trials. The number of items in the search display, the "set size", would be varied and the main dependent measure would be the reaction time or response time (RT), measured from the moment of stimulus onset until the moment that

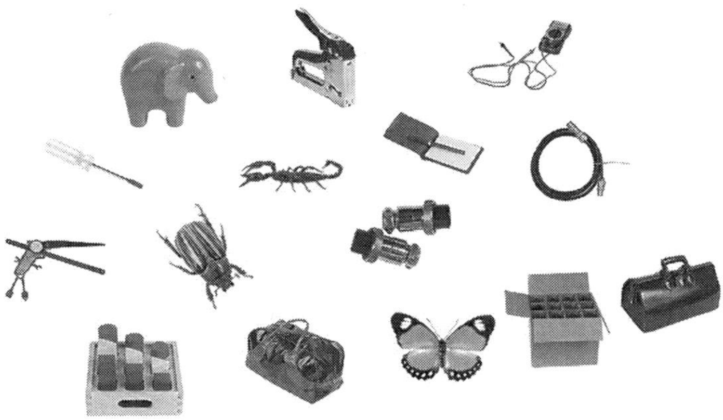

FIGURE 8.2 How many items per second can you process as you search this display for a beetle?

the key was pressed. Accuracy would also be recorded. In a task like the one in Figure 8.2, the results would look something like those shown in Figure 8.3.

The slope of the RT × set size function gives an estimate of the rate at which items can be processed in this search task. If one assumes that objects are being sampled without replacement from the display and that Os search exhaustively on the target absent trials (meaning that they examine every item before deciding that no target is present), then the 83 msec/item slope would imply a rate of about 12 items per second. Other assumptions are possible (e.g. sampling *with* replacement (Horowitz & Wolfe, 1998)) and these would lead to other estimates, but, for the data shown here, they would be in the range of about 12–25 items per second. Note that this means that it is *not* necessary to foveate each item. Fixational eye movements occur at a rate of 3–4 per second. When tasks are limited by the need to fixate each item (e.g. when items are very small), the RT × set size slopes become much steeper; on the order of 125 msec/item for target present and 250 msec/item for target absent. The data cartooned in Figure 8.3 come from stimuli where the set size was only 1–4 items and the items were deliberately large and well-separated to minimize "crowding" effects (Levi, 2008).

Problem Two

How can you search through objects so rapidly? An estimated rate of 12–25 objects per second raises a problem. If items are being processed one after another in series, this would mean that each item was being processed to the point of recognition in 40–80 msec. The problem is that, while a very brief exposure to a stimulus may suffice for identification (Potter et al., 2013), the act of identification continues after the stimulus is removed and estimates for the duration of

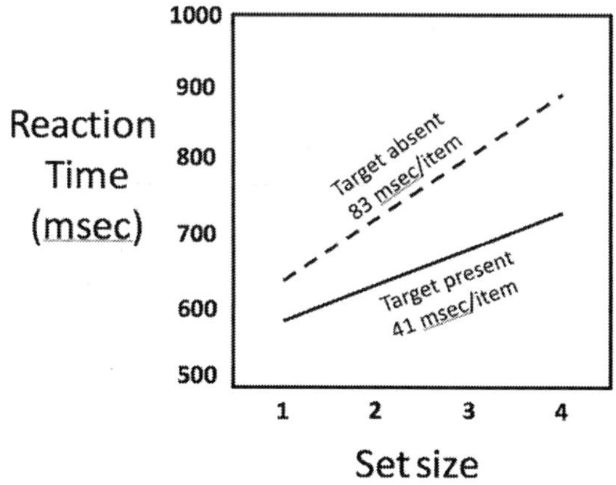

FIGURE 8.3 RT × set size functions for a search for an arbitrary object.

Redrawn from Wolfe et al (2011).

the object recognition process range from around 100 msec to several hundred msec (VanRullen & Thorpe, 2001) (Thorpe et al., 1996). For instance, if one takes the duration of the attentional blink as a measure of how long it takes to handle a single item, one would estimate 200–500 msec (Theeuwes et al., 2004; Ward et al., 1996). If both the overall rate of processing and the time to handle each item are more or less correct, it follows that more than one item must be processed at one time. One suggestion that has appeared multiple times over the years is that eye movements enforce some serial processing but that, once the eyes are fixated on some spot, all items within a functional field of view around that fixation are processed in parallel (Hulleman & Olivers, 2015). That parallel processing would need to be limited in its capacity. That is, it shouldn't be proposed that four items in the functional field of view (FFOV) can be processed as quickly as one. If that sort of parallel processing were possible, then RT × set size functions should be flat for small numbers of items. Obviously, that is not true for the data shown in Figure 8.3 and, in fact, it is not true in general. A limited capacity parallel story would work better. One could imagine that you have a full watering can to use on all the items within an FFOV. If there is only one item in the FFOV, you can dump the entire contents of your watering can on that item. If you have four items, you must distribute the resource more broadly.

There are various good reasons to propose that items are selected one at a time (Wolfe, 2003). If that is the case, we can modify the limited capacity parallel account to produce a model with serial and parallel aspects. A carwash provides a useful metaphor. Cars enter the carwash one at a time. However, at any given moment, multiple cars can be in the carwash, getting cleaned. Thus, if you looked at the input and output, you would see a serial process. If you dropped a large rock on the carwash, you would crush multiple cars in parallel. Using the timing of search, a car might go in every 40–80 msec, but it might take a few hundred msec to get out the other end (Wolfe, 2003). In computer science terms, this would be a "pipeline" architecture (Ramamoorthy & Li, 1977).

Problem Three

Even at that speed, how can you find anything in a real scene given how many objects might be in that scene? (And what is an "object" anyway?) For the present, let us agree that you can process 12–25 objects per second. That seems reasonably impressive. However, even at that rate, it is hard to see how we manage to find objects as quickly as we do in real scenes. For instance, in Figure 8.4, search for a bus.

That will not be a particularly difficult task. Indeed, it is surprisingly efficient given how many objects are in this scene. How many are there? This is actually an impossible question to answer definitively. Consider one of the taxi cabs, lined up on the right of the image. The cab is pretty clearly an object. What about a wheel or a window or the writing on the side of the door? Any of these could be the object of search so that suggests that they are objects in some sense. We could go on. What about shadows, arrows, etc? There is no absolute answer to this question (Neider & Zelinsky, 2008; Rosenholtz et al., 2007). As a consequence, there is no way to create definitive RT × set size functions for search in real scenes. That doesn't mean we can't try. We took advantage of the "LabelMe" database (Russell et al., 2008), a large set of scenes in which observers were asked to label all the objects. This effort has the problems, outlined above (no one labels shadows, for instance). However, the number of labels can be used as a conservative estimate

FIGURE 8.4 How many objects are in this scene?

By the way, that pyramid is the 2000-year-old tomb of Cestius in Rome, built when Egyptian artifacts were all the rage in Rome.

of the set size. When we performed a series of searches using the LabelMe count as the set size, we obtained very shallow RT × set size functions in the range of 4–8 msec/item. If we had a more comprehensive measure of the set size, the slopes would only become shallower. How can this be possible? How can we be processing items at a rate of 12–25/second in a random array of items like those in Figure 8.2 and have an estimate that is an order of magnitude faster when we are looking for an object in a scene like Figure 8.4?

An important part of the answer is "guidance" of attention. Under most circumstances, attention is not deployed randomly to objects. It is guided to some objects/locations over others by two broad forces: feature guidance and scene guidance. Feature guidance refers to guidance by properties of the target (and, to some extent, by properties of the distractors (Beck & Hollingworth, 2015)). Thus, in Figure 8.4, if you are asked to search for a bus, you will guide attention to horizontally oriented objects of approximate bus size. This will immediately limit your attention to a subset of objects that could plausibly be buses (Egeth et al., 1984). Notice that, in the case of three-dimensional scenes, guidance by size takes advantage of knowledge of the layout of the scene (Eckstein et al., 2017). Guidance by the features of the scene is limited to a relatively small subset of attributes. Color, size, orientation, and some more complex properties like "shininess" work. A property like "man-made object" would not guide attention (for a list, see Wolfe & Horowitz, 2017).

Feature guidance can explain the efficiency of many laboratory search tasks where the stimuli consist of arrays of items, placed at random in a two-dimensional display as in Figure 8.2. However, in the real world stimuli are not typically located at all locations with equal probability. If we return to the search for buses in Figure 8.4, you will note that the structure and content of the scene strongly constrain the possible location of buses. These scene guidance constraints can be divided into two varieties: syntactic and semantic (Biederman, 1977; Henderson & Ferreira, 2004; Vo & Wolfe, 2012). Buses do not float. Thus, it makes sense to look for them on horizontal surfaces that could support them against the force of gravity. This is a syntactic, structural, physical constraint. Buses belong on streets, not on sidewalks or the tops of walls. Semantic constraints of this sort embody our understanding of the scene and further limit where it makes sense to look for a bus.

Feature guidance can get going within a fraction of a second (Olds et al., 2000; Palmer et al., 2017). Interestingly, scene guidance is probably just about as fast though the precise experiments have not been done. Scene "gist" can be extracted from complex images after very brief exposures (Biederman et al., 1974; Greene & Oliva, 2009). This doesn't mean that you instantly know if the bus is present or where it is located. However, with a fraction of a second's exposure, you will know enough about the spatial layout of the scene to guide your attention. One approach to understanding the speed and effectiveness of scene guidance is to give observers a brief preview of a scene and then to allow the observer to see the display only in a narrow range around the point of fixation. (Castelhano & Henderson, 2007). That brief preview guides attention and the eyes quite effectively.

Problem Four

What happens if you are looking for more than one type of target at one time? If you are looking for a yellow Volkswagen, we now have some idea of what you might be doing. Somewhere in your mind/brain you have a representation of a yellow Volkswagen. The distinctive features of that representation will be used to guide attention. Thus, it seems likely that you would guide your attention toward yellow items in the scene (Egeth et al., 1984). Moreover, your understanding of the constraints on yellow Volkswagens would further constrain your search. But what happens if are looking for more than one type of target in a search? Suppose you are making dinner and you need the tomatoes, the basil, the mozzarella, and the black pepper. This would be an example of a "hybrid search". Hybrid search is a term, coined by Schneider and Shiffrin (1977) to describe tasks that involved a search of the visual world and a search of memory. In our example, in the course of search, your attention might be guided to the strawberries in the refrigerator. A memory search, through the set of four target items, would confirm that this was *not* a target and your attention would move onward to other candidate items. Actually, this would be an example of hybrid *foraging*. In foraging tasks, observers look for multiple instances of a target. For instance, picking blueberries is a foraging task in which observers search for blueberry after blueberry. Hybrid foraging is a visual search for multiple instances of multiple targets. In the present example, we can assume that all four ingredients need to be found before dinner can be prepared.

To get a feeling for how hybrid search works, memorize the four target objects in Figure 8.5 and then go back to Figure 8.2 and determine if any of those targets are present.

It is easy to memorize objects like those in Figure 8.5 even though, unlike our making-dinner example, these objects are unrelated to each other. Hundreds or thousands of such images can be encoded with only a few seconds exposure to each (Standing, 1973; Standing et al., 1970; Konkle et al., 2010a, 2010b). You would not be able to freely recall 100 images after seeing each for 3 seconds, but you would be very good at separating 200 images into 100 old and 100 new images on the basis of your ability to recognize the items in the memory set. Presumably, you found the volt meter in Figure 8.2 with little difficulty. In our formal version of the hybrid search task, we had observers search through displays of 1–16 items (the visual set size) for members of memory sets that varied in size from 1 to 100. Observers would memorize a set of items. We would test them, to make sure that the memory set was firmly lodged in memory and, then, they would be tested for several hundred trials, looking for those targets in visual displays that could vary in their set size.

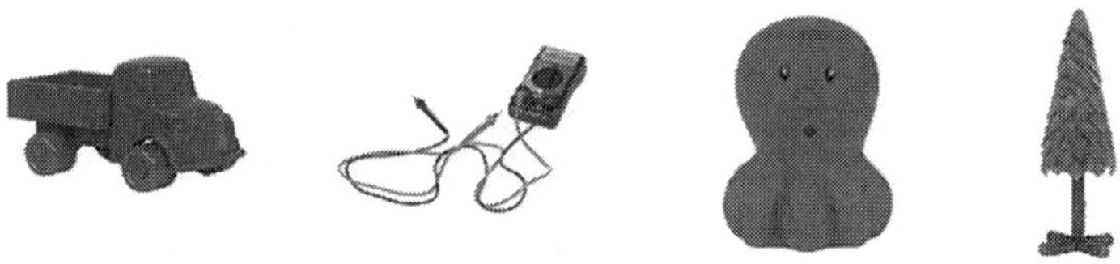

FIGURE 8.5 Are any of these items in Figure 8.2?

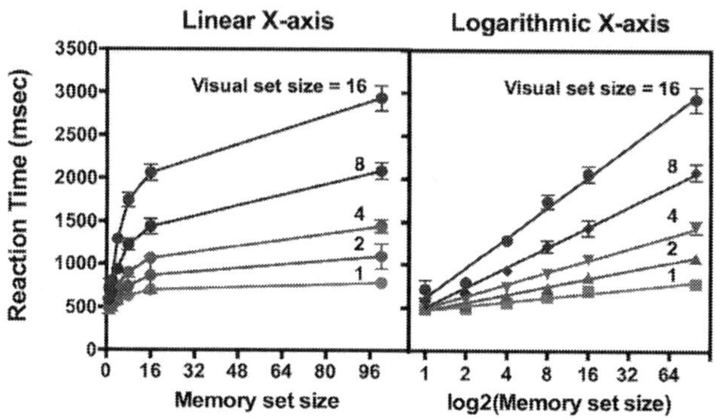

FIGURE 8.6 Basic hybrid search data redrawn from Wolfe (2012).

Even with a memory set size of 100, this is a very tractable task. RT increases with visual set size and with memory set size. The increase with visual set size is linear as is usually the case in visual search tasks. When you are holding just one item in memory, the slope of the RT × visual set size function is 15 msec/item. That slope rises to about 140 msec/item when the memory set size is 100. Unsurprisingly, it takes longer to decide that an item is or is not a target when there are 100 possible targets.

The pattern of results is different if we look at the data as a function of the memory set size. This is shown, plotted two ways, in Figure 8.6. The left side of the figure plots RT as a function of memory set size in the usual manner. It is clear that these are not linear RT × set size functions. On the right, the same data are plotted using a logarithmic x-axis. This makes it clear that RT rises linearly with the *log* of the memory set size. The log function is a remarkably good fit to the data. The regression lines in the figure are based on memory set sizes of 1–16 and then extrapolated to the data for the memory set size of 100. As can be seen, the fits of the lines are remarkably good, all the more remarkable because the memory set size of 100 was run with a new group of observers.

The logarithmic relationship between memory set size and RT saves us from some very laborious searches. If you had to search linearly through memory and the visual world, it would take many minutes to decide if any of your 1000 friends was present in a photo of, say, 50 people at a meeting. With the logarithmic compression of the memory search, that time is reduced to seconds. Of course, in the real world, you are not usually searching for one precise object in one precise position. You might be looking for scissors, tape, and wrapping paper without any specific commitment to the specific pair of scissors, roll of tape, or pattern of wrapping paper. It turns out that the rules of hybrid search are robust. Whether you are searching for one object in many poses (Alaoui-Soce et al., 2017) or object categories like "animal" (Cunningham & Wolfe, 2014) or words (Boettcher & Wolfe, 2015), the RTs are a linear function of visual set size and a logarithmic function of memory set size.

Why is the RT × memory set size function logarithmic? There are a number of possibilities. Here I sketch the answer from Leite and Ratcliff (2010). Let's imagine that information about the identity of an object accumulates over time in a noisy random walk: What Ratcliff and his colleagues have called a "diffusion" model of decision because information diffuses toward the decision boundary (Ratcliff, 1978; Ratcliff et al., 2016). In Figure 8.7A, there are three items in the memory set. One visual item is attended and we ask, "Is there evidence that this visual item matches one of the three items in memory." The answer is yes and, by time step 100, enough information has accumulated to generate a positive answer. Evidence for the other two items wanders around randomly. Now suppose we increase the number of items in memory to 18 (Fig 7B). Now information accumulates for each of these 18. Since random walks are random, there is a chance that they will wander too far. The more random walks, the greater is the chance that one will hit the decision boundary by mistake. In 8.7B, two of the random walks hit the bound, providing false evidence for the presence of other members of the memory set. To avoid this problem, one solution is to move the decision boundary up to a more conservative position. Raising the bound means that more evidence is required before committing to a positive decision. This

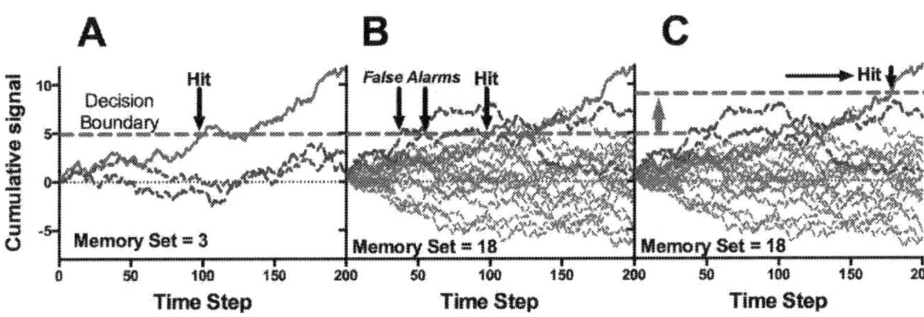

FIGURE 8.7 Cartoon of the Leite and Ratcliff (2010) account of logarithm memory search.

(A) Information accumulates in a noisy fashion toward a decision bound. (B) With many items in the memory set, there is an increased chance that a distractor will reach the decision bound. So (C) the bound can be raised to avoid that problem. The resulting increase in RT is a log function of the memory set.

is shown in 8.7C. If you require more evidence, you are going to require more time. In this cartoon, the RT moves out to 175 time steps, but the false alarms are avoided. Leite and Ratcliff show that, at least for their class of diffusion model, if the boundary is moved to keep the false alarm rate constant, the increase in RT will be a log function of the number of diffusion processes.

Problem Five

Where is that yellow Volkswagen detector? Suppose that you were looking for members of the following set of items: a yellow Volkswagen, a lemon, a yellow duckling, and a yellow traffic sign. Once you attended to one of the items in the visual display, the foregoing account of hybrid search would say that you would launch four diffusion processes, one for each of the items in the display. If accumulating information reached the yellow Volkswagen decision boundary, we would declare that we had found a yellow Volkswagen. That diffusion processes could be thought of as a yellow Volkswagen detector. If we unpack this a bit more, it seems that there must actually be two acts of detection that are occurring here. Return to Figure 8.2 and look for any of the four targets: yellow Volkswagen, lemon, duckling, or sign. To begin, you will select an item from the visual display. Since you are looking for a yellow Volkswagen, we can imagine that your attention will be *guided* to yellow items. Let us suppose guidance directs your attention to the yellow screwdriver. In step one, you identify this as a yellow screwdriver. In step 2, you determine that a yellow screwdriver is not what you are searching for and you will then move to the next item and repeat.

Why two steps? Why is it not adequate to say that a single step determines that this is a screwdriver and, thus, not a car, lemon, duckling, or sign? Even though it is not part of the target set, you clearly identify the screwdriver. Somehow, you compare the visual item to all the objects in long term memory and come up with the identity, "screwdriver". In the same way, you can look at the scene around you at the moment and identify objects though you have no particular set of objects in mind. Clearly, the basic object identification task is not dependent on the number of items in the current memory set. Thus, in a hybrid search task, observers must identify the object and then check that object against the memory set. It is that second test whose RT is logarithmically dependent on the memory set size.

The representations of the search targets are the search "templates", mentioned earlier in this chapter. It is often proposed that they are resident in working memory (Olivers et al., 2011). There is a large body of work showing interactions between working memory and visual search. For example, holding the color "red" in working memory will tend to bias subsequent visual search toward red items. However, working memory cannot be the home of the memory set in hybrid search. There is much debate about the details of working memory (Suchow et al., 2014) but everyone agrees that it has a small capacity. Working memory holds the equivalent of about four items (Cowan, 2001). One can argue that working memory is a resource that can be spread over more than four items (Bays & Husain, 2008) or that only one item can serve as the search template at one time (Olivers et al., 2011), but no one can argue that 100 items can be held as search templates in working memory. Moreover, we have done multiple experiments where we loaded working memory and then had observers perform a hybrid search task. Hybrid search was not impaired when working memory was "full" (Drew et al., 2015). We prefer to think of the memory set as being held in "activated long term memory (ALTM)" (Cowan, 1995), a hypothetical memory structure that holds the elements in long term memory that are relevant for the current task.

This leaves us with a problem. As noted, we know that working memory and visual search interact. If working memory is not holding the yellow Volkswagen detector during search, what is it doing? The answer, we believe, is that it is holding the yellow Volkswagen, itself. Well, not the actual object, of course; working memory may be holding the visual representation of the current object of attention, a yellow Volkswagen or whatever other object it may be. After all, the representations in memory are not being compared directly to the physical object in the world. They must be compared to some representation of the current object of attention and we suggest that this representation, built in the visual system, is held in working memory. This is a theory that could benefit from much more empirical support. There is one nugget of evidence in the Drew et al. (2015) experiments in which loading working memory did not influence hybrid search. In those experiments, while working memory did not influence hybrid search, hybrid search did influence working memory. Specifically, doing a hybrid search task seemed to use up one "slot" in working memory. If you could reliably recall four colored boxes in a pure working memory task, you could recall only three when a hybrid search task intervened between encoding and recall. It was as if attending to item after item in the visual display required a dedicated piece of working memory; one "slot" to hold the current object of attention.

Thus, pending more research, here is an account of hybrid search.

1) An item is selected as the current object of attention. That selection is not random; at least, not if there is a possibility of some useful guidance. In the example given above, attention would tend to be directed to yellow objects if all target items are known to be yellow.
2) The selected item is represented in working memory.
3) That working memory representation is somehow compared against all the contents of long term memory and identified. If the target set consisted of eight animals and the object was identified as a fork, the item could be rejected at this stage.
4) If the item could be a member of the memory set of targets, then the memory set must be searched and a decision reached. Is this or is this not a target?

5) If it is not a target (or if more than one target can be present in the visual display (Wolfe et al., 2016)), attention selects a new object and the process repeats.

If a yellow Volkswagen is a candidate target and if it is the current object of attention, it will be represented three times: once as a visual stimulus, held in working memory; once as a long term memory representation, there to identify yellow Volkswagens any time they appear, and once in activated long term memory, as a member of the currently relevant memory set. The working memory and ALTM representations might be two ways of describing a single neural representation but, in the cognitive architecture the yellow Volkswagen memory gets consulted in two ways.

I am sorry that Naomi Weisstein is not with us anymore so I can't ask her what she thinks about all of this. I like to think she would be glad that her yellow Volkswagen detectors were provoking new thought and new experiments. And I am sure that, if she thought it was nonsense, she would have let me know.

Note

1 To see color versions of the figures in Chapter 8 go to https://research.franklin.uga.edu/visual-perception-laboratory/chapter-8

References

Alaoui-Soce, A., Zou, B., & Wolfe, J. M. (2017). *Variable viewpoint hybrid search: Searching for the object or the image?* Paper presented at the Annual Meeting of the Vision Science Society, May 2017, abstract 23.3006.

Barlow, H. (1995). The neuron doctrine in perception. In M. S. Gazzaniga (Ed.), *The Cognitive Neurosciences* (pp. 415–435). Cambridge, MA: MIT Press.

Bays, P. M., & Husain, M. (2008). Dynamic shifts of limited working memory resources in human vision. *Science, 321*(5890), 851–854.

Beck, V. M., & Hollingworth, A. (2015). Evidence for negative feature guidance in visual search is explained by spatial recoding. *Journal of Experimental Psychology: Human Perception and Performance, 41*(5), 1190–1196.

Biederman, I. (1977). *On processing information from a glance at a scene: Some implications for a syntax and semantics of visual processing.* Paper presented at the Proceedings of the ACM/SIGGRAPH Workshop on User-oriented Design of Interactive Graphics Systems, Pittsburgh, PA.

Biederman, I., Rabinowitz, J. C., Glass, A. L., & Stacy, E. W. Jr., (1974). On the information extracted from a glance at a scene. *Journal of Experimental Psychology, 103*(3), 597–600.

Boettcher, S., & Wolfe, J. M. (2015). Searching for the right word: Hybrid visual and memory search for words. *Attention, Perception & Psychophysics*, 77(4), 1132–1142.

Castelhano, M. S., & Henderson, J. M. (2007). Initial scene representations facilitate eye movement guidance in visual search. *Journal of Experimental Psychology: Human Perception and Performance, 33*(4), 753–763.

Cowan, N. (1995). *Attention and Memory: An Integrated Framework.* New York: Oxford University Press.

Cowan, N. (2001). The magical number 4 in short-term memory: a reconsideration of mental storage capacity. *Behavioral & Brain Sciences, 24*(1), 87–114; discussion 114–185.

Cunningham, C. A., & Wolfe, J. M. (2014). The role of object categories in hybrid visual and memory search. *Journal of Experimental Psychology: General, 143*(4), 1585–1599.

Drew, T., Boettcher, S. P., & Wolfe, J. M. (2015). Searching while loaded: Visual working memory does not interfere with hybrid search efficiency but hybrid search uses working memory capacity. *Psychonomic Bulletin & Review, 23*(1), 201–212.

Eckstein, M., Koehler, K., Welbourne, L. E., & Akbas, E. P. (2017). Humans but not deep neural networks often miss giant targets in scenes. *Current Biology, 27*(18), 2827–2832.

Egeth, H. E., Virzi, R. A., & Garbart, H. (1984). Searching for conjunctively defined targets. *Journal of Experimental Psychology: Human Perception and Performance, 10*, 32–39.

Greene, M. R., & Oliva, A. (2009). The briefest of glances: the time course of natural scene understanding. *Psychological Science, 20*(4), 464–472.

Gross, C. G., Rocha-Miranda, C. E., & Bender, D. B. (1972). Visual properties of neurons in inferotemporal cortex of the macaque. *Journal of Neurophysiology, 35*, 96–111.

Henderson, J. M., & Ferreira, F. (2004). Scene perception for psycholinguists. In J. M. Henderson & F. Ferreira (Eds), *The Interface of Language, Vision, and Action: Eye Movements and the Visual World* (pp. 1–58). New York: Psychology Press.

Horowitz, T. S., & Wolfe, J. M. (1998). Visual search has no memory. *Nature, 394*(6693), 575–577.

Hulleman, J., & Olivers, C. N. L. (2015). The impending demise of the item in visual search. *Behavioral and Brain Sciences, 40*, 1–76.

Konkle, T., Brady, T. F., Alvarez, G. A., & Oliva, A. (2010a). Scene memory is more detailed than you think: The role of categories in visual long-term memory. *Psychological Science, 21*, 1551–1556.

Konkle, T., Brady, T. F., Alvarez, G. A., & Oliva, A. (2010b). Conceptual distinctiveness supports detailed visual long-term memory for real-world objects. *Journal of Experimental Psychology: General, 139*(3), 558–578.

Krizhevsky, A., Sutskever, I., & Hinton, G. E. (2012). ImageNet classification with deep convolutional neural networks. *Advances in Neural Information Processing Systems, 25*(2), 1097–1105.

Leite, F. P., & Ratcliff, R. (2010). Modeling reaction time and accuracy of multiple-alternative decisions. *Attention, Perception & Psychophysics*, 72(1), 246–273.

Lettvin, J. (1995). J.Y. Lettvin on grandmother cells. In M. S. Gazzaniga (Ed.), *The Cognitive Neurosciences* (pp. 434–435). Cambridge, MA: MIT Press.

Levi, D. M. (2008). Crowding–An essential bottleneck for object recognition: A mini-review. *Vision Research, 48*(5), 635–654.

Neider, M. B., & Zelinsky, G. J. (2008). Exploring set size effects in scenes: Identifying the objects of search. *Visual Cognition, 16*(1), 1–10.

Olds, E. S., Cowan, W. B., & Jolicoeur, P. (2000). The time-course of pop-out search. *Vision Research, 40*(8), 891–912.

Olivers, C. N., Peters, J., Houtkamp, R., & Roelfsema, P. R. (2011). Different states in visual working memory: When it guides attention and when it does not. *Trends in Cognitive Sciences, 15*(7), 327–334.

Palmer, E. M., Van Wert, M. J., Horowitz, T. S., & Wolfe, J. M. (2017). Measuring the time course of selection during visual search. *Attention, Perception & Psychophysics*, submitted Aug 2017.

Potter, M. C., Wyble, B., Hagmann, C. E. H., & McCourt, E. (2013). Detecting meaning in RSVP at 13 ms per picture. *Attention, Perception & Psychophysics, 76*(2), 270–279.

Ramamoorthy, C. V., & Li, H. F. (1977). Pipelined Architecture. *Computing Surveys*, *9*(1), 61–102.

Ratcliff, R. (1978). A theory of memory retrieval. *Psychological Review*, *85*(2), 59–108.

Ratcliff, R., Smith, P. L., Brown, S. D., & McKoon, G. (2016). Diffusion decision model: Current issues and history. *Trends in Cognitive Sciences*, *20*(4), 260–281.

Rosenholtz, R., Li, Y., & Nakano, L. (2007). Measuring visual clutter. *Journal of Vision*, 7(2), 1–22.

Russell, B. C., Torralba, A., Murphy, K. P., & Freeman, W. T. (2008). LabelMe: a database and web-based tool for image annotation. *International Journal of Computer Vision*, 77(1–3), 157–173.

Schneider, W., & Shiffrin, R. M. (1977). Controlled and automatic human information processing: I. Detection, search, and attention. *Psychological Review*, *84*, 1–66.

Standing, L. (1973). Learning 10,000 pictures. *Quarterly Journal of Experimental Psychology*, *25*, 207–222.

Standing, L., Conezio, J., & Haber, R. N. (1970). Perception and memory for pictures: Single trial learning of 2500 visual stimuli. *Psychonomic Science*, *19*, 73–74.

Suchow, J. W., Fougnie, D., Brady, T. F., & Alvarez, G. A. (2014). Terms of the debate on the format and structure of visual memory. *Attention, Perception & Psychophysics*, *76*(7), 2071–2079.

Theeuwes, J., Godijn, R., & Pratt, J. (2004). A new estimation of the duration of attentional dwell time. *Psychonomic Bulletin & Review*, *11*(1), 60–64.

Thorpe, S., Fize, D., & Marlot, C. (1996). Speed of processing in the human visual system. *Nature*, *381*(6582), 520–552.

Treisman, A., & Gelade, G. (1980). A feature-integration theory of attention. *Cognitive Psychology*, *12*, 97–136.

VanRullen, R., & Thorpe, S. J. (2001). Is it a bird? Is it a plane? Ultra-rapid visual categorisation of natural and artifactual objects. *Perception*, *30*(6), 655–668.

Vo, M. L., & Wolfe, J. M. (2012). *ERPs dissociate semantic and syntactic processing in scenes*. Paper presented at the 2012 European Conference on Visual Perception (ECVP), Alghero, Italy.

Ward, R., Duncan, J., & Shapiro, K. (1996). The slow time-course of visual attention. *Cognitive Psychology*, *30*(1), 79–109.

Weisstein, N., & Harris, C. S. (1974). Visual detection of line segments: An object superiority effect. *Science*, *186*, 752–755.

Wolfe, J. M. (2003). Moving towards solutions to some enduring controversies in visual search. *Trends in Cognitive Science*, 7(2), 70–76.

Wolfe, J. M. (2012). Saved by a log: How do humans perform hybrid visual and memory search? *Psychological Science*, *23*(7), 698–703.

Wolfe, J. M., & Horowitz, T. S. (2017). Five factors that guide attention in visual search. [Review Article]. *Nature Human Behaviour*, *1*, 0058.

Wolfe, J. M., Palmer, E. M., & Horowitz, T. S. (2010). Reaction time distributions constrain models of visual search. *Vision Research*, *50*, 1304–1311.

Wolfe, J. M., Aizenman, A. M., Boettcher, S. E. P., & Cain, M. S. (2016). Hybrid Foraging Search: Searching for multiple instances of multiple types of target. *Vision Research*, *119*, 50–59.

Wolfe, J. M., Alvarez, G. A., Rosenholtz, R., Kuzmova, Y. I., & Sherman, A. M. (2011). Visual search for arbitrary objects in real scenes. *Attention, Perception & Psychophysics*, *73*(6), 1650–1671.

9

WHEN VISUAL ATTENTION HURTS

Elisabeth M. Fine and Adam Reeves
Northeastern University

Introduction

We tend to think of attention as a unitary concept, of a somewhat similar level of abstraction as memory or perception; we recall paying attention to something we saw, while thinking little of the stages or processes involved. However, a scientific study of attention, while it should keep in mind that attention serves to make information salient, must penetrate beneath the surface appearance to locate the sub-processes necessary for it to operate. This has turned out to be hard, as attention can be defined in ways that are not always congruent. To illustrate, we can define attention as a *task* (the canonical instructions being, 'Pay attention' or 'Be vigilant' in case of rare events), or as a *mental construct* (e.g., concentration, focusing, effort, activation, selection, synthesis, feature integration), or as a *behavioral construct* (an attended stimulus is one that controls a specified behavior), or as a *neurological construct* (a change in single cell response or in gross recording like the EEG, VEP, or BOLD response), or as a *processing flow* (top-down: endogenous or feedback, versus bottom-up: exogenous or feed-forward), or as a *processing mode* (parallel, serial, or hybrid), or as an *individual difference* (being attentive or distractible; global or analytic) or as a *philosophical construct* (choice or free will; awareness or consciousness), or as a form of *explanation*, as in, 'I didn't follow your argument at first because I wasn't paying attention.' Given this mélange, it is hardly surprising that attention research is far from unitary. So, how to proceed? The experimental psychologist needs suitable methods of measurement, as sketched in Table 9.1. Attention can be measured as a *direct effect* of an instruction (e.g., 'Pay attention'; 'concentrate'; 'split your attention 80:20'), or as an *inference* from a trade-off (a loss on task A when task B is concurrent), or as higher *sensitivity* to an attended stimulus, or as *feature integration*, or from *search rate*, or as an outcome of *selective*

TABLE 9.1 Varieties of attention

Definition	*Method*	*Example*
task command/instruction	'Pay attention' to A versus B	trade-off
concentration (focus)	cue to focus vs. neutral cue	probe-signal
effort	increase/decrease motivation	fatigue
vigilance	attend to rare events	Mackworth clock
analysis or selection	pick out targets from distractors	visual search
synthesis or integration	feature integration (Triesman)	conjunction
cathexis: unconscious attention	object attachment	association
behavioral control	reinforcement	extinction
sharpening neural profile	conditioning/recording	single cell
steady-state potential	component analysis of EEG	hemisphere difference
evoked potential	waveform of VEP, AEP	Stimulus locking
processing flow	top-down/bottom up	instruction
processing mode	serial/parallel/hybrid	modeling latencies
attention deficit	individual difference	neuropsychology
processing style	global/analytic personality type	questionnaire
free will	choice	test limits of choice
awareness	self report; brain activity	altered states
theater of attention	theoretical synthesis	limited capacity
central processing unit	working memory	recall/interference
naïve explanation	self report; self exculpation	structured interviews

reinforcement (Dinsmoor, 1985), or from *cell responses* (a sharpened receptive field or a speeding up) or from *individual differences* (e.g., attention deficit disorder.) Hence there are at least as many different methods for measuring attention as there are outcomes of attention. In each case, if attention changes behavior relative to an appropriate control, one may further ask whether this is due to changes in sensitivity or in the criterion? If a change in sensitivity, is this from enhancing targets, from increasing overall activation, from selecting a sub-set of items, or from suppressing distractors (Lu & Dosher, 1999)? If to a change in criterion, is this related to perceived pay-offs, and is the change optimal, say in a Baysian sense? Finally, does any change due to attention reflect the outcome of a parallel, a serial, or a hybrid mechanism?

Our purpose is not just to list the 'varieties of attention' but also to point out that a common intuition about attention may lead one astray. Intuition suggests that paying attention enhances perception; indeed, it has been taken as a defining characteristic that items selected by attention will be processed better, and rejected items worse, than neutral items (Posner et al., 1978; for corresponding neural data, see Kanwisher & Wojciulik, 2000.) The Skinnerian definition of attention as referring to the stimulus that 'controls' behavior, although neat, also implies this: accuracy, as defined by stimulus-response probabilities, *must* increase if the stimulus is attended (i.e., more effectively controls the response.) Indeed, consistent with this intuition, empirical evidence that selective attention enhances perception has been obtained in several experimental designs. In a *block design with foreknowledge*, subjects are told in advance where (or how) a stimulus may appear in a block of trials, so that they may focus on it more effectively (Grindley & Townsend, 1968). For example, in a probe-signal experiment, a 'signal' tone presented at a frequency expected throughout a trial block is detected more reliably than a 'probe' tone presented at an unexpected frequency (Greenberg & Larkin, 1968), and line segments shown at an expected orientation ('signals') are detected more accurately than line segments ('probes') presented at an unexpected orientation (Kurylo et al., 1996). In a *randomized design with cueing*, a cue is presented before the test stimulus, with the cues randomized across trials within each block. Cued letters are reported more accurately than uncued letters in an experiment in which letters are presented in a ring around fixation to equalize acuity (Eriksen & Yeh, 1985), with letters adjacent to and opposite to the cued letter being faster than those in-between (Skelton & Eriksen, 1976; Bahcall & Kowler, 1999). Moreover, the time available to focus attention in this design can be measured by varying the cue-signal delay. For example, thresholds for detecting an unexpected tone in noise improve 52 ms after a cue is presented, suggesting that focusing takes less than 1/20th of a second (Scharf et al., 2007). The positive effects of attention can transcend location; e.g., in an *object-cueing* procedure, attended objects are typically processed more precisely than un-attended ones (Egly et al., 1994; Blaser et al., 2000), even when the locations of attended and unattended objects are identical.

Given all this evidence that attention improves performance, it seems natural to ask, why? Recent signal-detection inspired models suppose that attention filters out noise, decreases uncertainty about the signal (Lu & Dosher, 1999, 2008), suppresses irrelevant channels while leaving the signal unchanged (Scharf et al., 1997), or directly enhances the signal (Carrasco et al., 2002). In each case, attention has been shown to improve performance (d') (equivalent to improving accuracy with a criterion-free method). A focus on signal detection, rather than on broader organismic factors, seems justified when subjects are appropriately motivated; that is, not distracted, indifferent, lethargic, unclear about the task, overly aroused, or abnormal. Since attention to the wrong location or the wrong spatial frequency band impairs performance even when the subject is motivated (Yeshurun & Carrasco, 1998), the generalization that attention improves performance also presupposes *validity*: that is, (a) attention is paid to the signal rather than to irrelevant or competing locations or features, (b) the task instructions in the blocked design and cuing procedure in the randomized design are understood and are not misleading.

Given appropriate motivation and valid instructions and cues, is the generalization that attention aids performance still always correct? In the following research we restricted our experiments to those in which *the signal is known in advance, its form unvarying*, and *its spatial location is validly cued*, and we ask whether attention necessarily improves processing in this restricted domain. We employed *criterion free* methods so that any variation of accuracy with attention had to be due to variations in processing, not uncontrolled variations in response criteria. We also needed to narrow down our choice of 'attentional' variable, given the aforementioned plethora of meanings associated with the term. We chose *selective attention* as the primary meaning (James, 1890), and ran conditions in which selective attention was endogenous (top-down) rather than exogenous (bottom-up). In the latter case, we reasoned that, to the extent that so-called 'salient' stimuli are just those which automatically attract attention, greater salience must, by definition, improve processing. However, in the endogenous case, there is no certainty that top-down attention must improve processing, despite intuition and despite current models, since, as we show below, signal detection models can in fact predict that performance improves, does not change, or declines, when attention increases. We also chose conditions in which processing is parallel rather than serial (Najemnik & Geisler, 2008; Eckstein, 1998; Santhi & Reeves, 2004; Nakayama & Bravo, 1992). Independent of the model chosen, however, our experiment illustrates a fundamental test of the hypothesis that increased attention improves performance, as it leaves no 'wiggle room' for alternative explanations in terms of signal certainty, criteria, validity, or salience. If increasing attention does reduce performance in this case, it may well do so in other situations.

In Eriksen's classic study (Eriksen & Yeh, 1985), although attention was under endogenous (task-driven) control, there were also exogenous (stimulus-driven) influences due to letter onsets. We therefore employed the 'offset' method of Jonides and Yantis (1988) in an attempt to ensure that attention was endogenous. Participants viewed a screen on which four 'placeholders' appeared at top, bottom, right, and left of fixation. Each placeholder was a rectangular box incorporating horizontal, vertical, and diagonal lines. After a short period, one of the placeholders changed into a letter by turning off some of the line segments which formed it. Thus our letters would not attract attention exogenously, since only onsets, not offsets, appear to force attention independent of instructions (Jonides & Yantis 1988; Donaldson & Yamamoto, 2016; Owens & Spehar, 2016).

General Methods

Subjects were young adults (18 to 30 years old) with normal visual acuity (at least 20:20 in each eye) enrolled in introductory courses at Northeastern University. They earned credit for their one-hour participation. They were free to leave the experiment at any time without losing credit.

Apparatus. Stimuli were presented on a 17 inch. Sony Multiscan 200ES 83 Hz raster-based monitor under the control of a MAC G3 office computer running OS9 and custom code employing the RSVP software package created by Michael

Tarr of Brown University. Screen luminance was 10 cd/m^2. Stimuli were black line segments shown at maximum contrast. The screen was viewed at 40 cm using a head rest and subtended 34° by 43° at the eye. Room lighting was dim. Stimulus presentation was accurate to one frame (12ms) as confirmed with a photodiode and oscilloscope.

Stimuli were letters, Es, or box-like placeholders shown around a central fixation cross (Figure 9.1), which subtended 0.57 deg. at the eye when presented at 5 deg. eccentricity and 1.02 deg. at the eye when presented at 10 deg. eccentricity. Offset stimuli were created as in Jonides and Yantis (1988) by removing segments from placeholders as needed to form identifiable letters. In the E-discrimination conditions, removal of the two right-hand vertical segments of the box revealed a forward-pointing E, and removal of the two left-hand vertical segments, a backward-pointing E. In the letter-identification conditions, different offset letters were formed by removing segments from the placeholder as needed.

Procedure. Trials began with a 1025ms fixation screen (see Figure 9.1). Four placeholders were then presented for 250 ms, and followed by an offset target for 100 ms (letter) or 50 ms (E). Targets were created by turning off line segments in a randomly-chosen placeholder. The target was then replaced by a placeholder, so that for the last 250ms on each trial, the subject viewed the same four placeholders as at the start. These after-coming placeholders acted as an effective backward patterned mask (Turvey, 1973), being similar in shape to the target stimuli (Haber, 1970). The target letter was identified by clicking on a fixed post-trial display of 20 letters arranged as a 4×5 matrix (not shown in Figure 1). In different

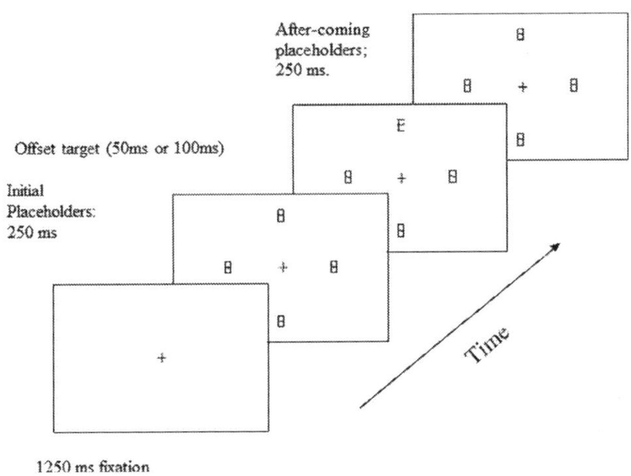

FIGURE 9.1 Display sequence.

Stimulus durations for offset targets were 50ms for Es and 100ms for letters. Initial and final placeholders were always 250ms.

experiments, target letters were either selected from one of two possible letters (T or B for half the subjects; B or H for the rest), or from one of 10 or 20 possible letters. The orientation of the E was reported by pressing F for a backward E and J for a forward E on a keyboard in which those two letters had raised bumps to guide the fingers. Subjects self-initiated trials by pressing the space bar, and were free to stop the experiment if they needed to. They were told before every experiment that conditions were difficult, so they would need to concentrate, and even then, they would not always be correct. In post-session de-briefing we explained that we were measuring the effect of focusing attention on perception in difficult conditions.

Experiment 1. Letters in Cardinal Locations

Speed and accuracy were measured for identifying offset letters presented in the four cardinal locations, top, bottom, left, and right of fixation. Attention was either *diffused* to all four locations or *focused* on two opposite ones, top-bottom or left-right, according to instructions that were always valid. Since trials were blocked in this and subsequent experiments, it was possible for the subject to continuously monitor all four positions when attention was diffused, or monitor the two chosen positions when attention was focused, throughout a block of trials; switching strategies was not required.

Subjects

Subjects were 27 Northeastern undergraduates from the subject pool, whose acuity was at least 20/20 in each eye. Two subjects with incomplete data were replaced.

Stimuli and Procedure

The box-like placeholders and offset letters were shown at 5 deg. eccentricity, at top, bottom, left, and right of central fixation (see General Methods, and Figure 9.1). The subject was required to click on the target letter (B or H for half the subjects; V or T for the rest) in the post-display array that they thought had been presented, making their best guess if uncertain; they were asked to respond quickly without sacrificing accuracy. Subjects were informed verbally before each block of 24 trials that one offset letter would appear on each trial, either in top-bottom, or left-right, or any of the four locations. Block order was randomized across subjects. Focusing was blocked: half the subjects monitored all four locations or top and bottom; the other half monitored all four locations or left and right. They practiced for 12–15 minutes before data were collected. Sessions lasted 50 to 65 minutes.

Results

Mean accuracy was 74.2 percent when two locations were monitored and 76.9 percent when four were monitored. This difference of 2.7 percent in favor of monitoring 4 exceeded the standard error of 1.55 percent, but not significantly ($p = 0.15$). An ANOVA with replications applied to the arc-sine transformed percentage correct data yielded a highly significant triple interaction ($F_{3,78} = 20.65$, $p < 0.001$), however, in which accuracy on the top and bottom locations averaged 10 percent better, but accuracy on the left and right locations averaged 5 percent worse, when monitoring 4 locations than 2 locations. Figure 9.2 visualizes this interaction; bars represent +1 standard error. Focusing may have aided the left-right locations slightly, but clearly disrupted the top-bottom ones.

Latencies were determined from the means of the subject's median RTs in each condition, in this and subsequent experiments. (Median RTs were used so that no trials needed to be eliminated as outliers.) Latency data showed that focusing on 2 locations was not only less accurate, but also slower than diffusing attention to all 4 locations, by 24 ms on average ($F_{1,26} = 12.5$, $p < 0.01$). This main effect was due not to the left-right locations, which differed by less than one ms, but to the top-bottom ones, where monitoring two locations was slower by 49 ms than monitoring all four (interaction $F_{3,26} = 7.78$, $p < 0.01$), as shown in Figure 9.3. The conclusion that *focusing both lowers accuracy and slows reaction times for the top-bottom* locations, *but has little effect on the left-right ones*, was also true of both the 10 most and 10 least accurate subjects when their data were separated out prior to

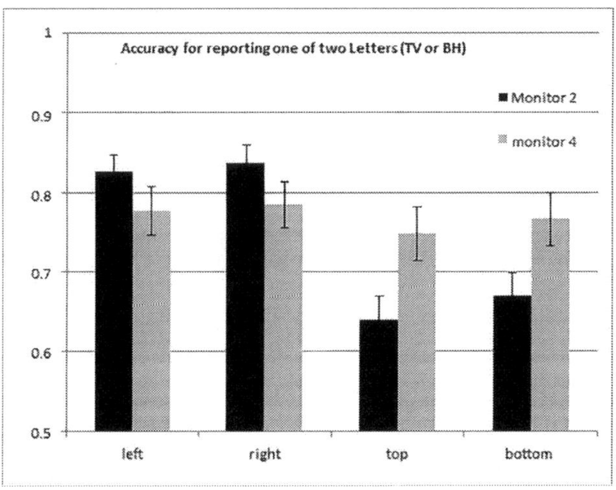

FIGURE 9.2

Accuracy for reporting one of two possible offset target letters as a function of location in the display (top, bottom, left, or right) and focusing (monitoring 2 or 4 locations), in Experiment 1. Bars given ±1 standard error.

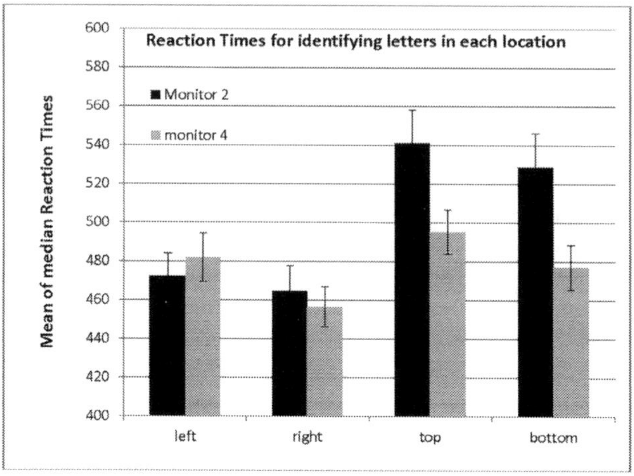

FIGURE 9.3

Latency for reporting one of two possible offset target letters, plotted against location of the target letter, when monitoring two or four positions, from Experiment 1. Bars give ±1 standard error.

applying ANOVA, so this is unlikely to be the result of aberrant performance by untypical subjects.

Further results

We have run additional experiments with letters to compare results at different eccentricities, 5 or 10 deg. (Fine & Reeves, 2001), using 12 to 15 different subjects in each case. Focusing on the top and bottom locations was significantly inferior to diffusing attention to all four locations at both eccentricities. We conclude that the choice of location in the periphery may not be germane to the effect; however, we cannot conclude that eccentricity is entirely irrelevant as we have not run comparable experiments in the fovea or in the extreme periphery.

Experiment 2. Letters in Diagonal Positions

Experiment 1 found that monitoring the top and bottom locations for a letter was inferior to monitoring all four locations, whereas there was no effect of monitoring on left and right locations. Experiment 2 was conducted to determine if the difficulty in monitoring the top and bottom locations was confined to the midline, as might be inferred from Hughes and Zimba (1985), or reflected some more general property of the focusing of spatial attention (see e.g., Klein & McCormick, 1989). We therefore re-ran the experiment using a diagonal format, in which stimuli were top-right, top-left, bottom-right, and bottom-left. When focusing on just two locations, these were either top-right and bottom-left, or top-left and bottom-right.

Procedure

As in Experiment 1, but the stimulus locations were rearranged with placeholders at top-right, top-left, bottom-right, and bottom left, at 5 deg. eccentricity. As before, a randomly chosen placeholder was turned into a random letter for 100 ms by removing appropriate line segments, before turning back into a placeholder for a further 250 ms. The 15 subjects reported the letter identity using the same final screen as in Experiment 1. Unlike Experiment 1, in which the target was one of two letters, in Experiment 2 the target letter could be any one of 20 letters. The task of monitoring 2 or 4 locations was blocked, as before.

Results

Figure 9.4 plots mean accuracy for reporting letter identity against location. Accuracy when all four locations were monitored averaged 56 percent, significantly higher than when only two were monitored (52 percent) ($F_{1,14} = 9.91, p < 0.01$). The effect was marginally larger at upper left than at lower right (interaction $F_{3,42} = 2.23, p = 0.099$). Figure 9.5 plots the corresponding latencies. Responses when all four locations were monitored were significantly faster than when two locations were monitored ($F_{1,14} = 12.5, p < 0.01$), especially on the right side as shown by a significant interaction of number monitored with location ($F_{3,44} = 3.56$, $p < .05$). Although focusing attention on two locations always made the task more difficult, a comparison of Figures 4 and 5 shows that focusing sacrificed speed, but not accuracy, on the right side and accuracy, but not speed, on the left side. We do not know why this happened, and it may be spurious given the small sample size,

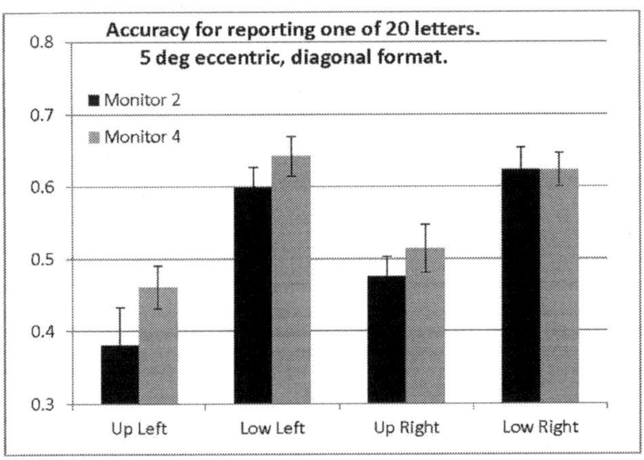

FIGURE 9.4

Accuracy for reporting one of 20 offset letters presented diagonally at 5° eccentric, plotted against the location of the target letter, in Experiment 2. Accuracy in monitor 4 was higher than in monitor 2.

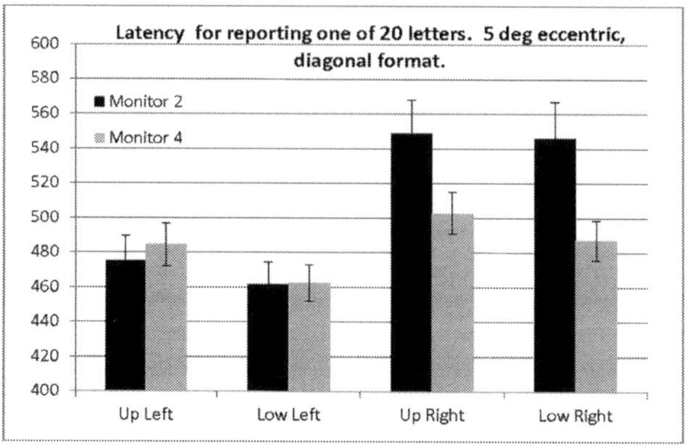

FIGURE 9.5

Latency for reporting one of 20 possible offset letters in the diagonal format, in Experiment 2. Latency was faster when monitoring 4 rather than 2 locations, but only on the right.

but at least we can conclude from the main effects that focusing on two locations hurts performance as compared to diffusing attention to all four locations. This conclusion agrees with that from Experiment 1 and indicates that the mid-line is not the critical factor in the poorer identification of masked offset letters under focusing.

Experiment 3: E Discrimination

We repeated Experiment 1 using optotypes, namely, backward- and forward-pointing offset Es. These can be discriminated using low-level features, specifically orientation, and do not require the complex feature analysis needed to identify individual letters. We anticipated that focusing attention might aid detection of the Es, or perhaps have no effect, but would not harm them, given our earlier finding of the beneficial effects of attention on orientation discrimination of briefly-presented masked line segments (Kurylo et al., 1996). However, we had no strong expectation, since (to our knowledge) the numerous previous studies which have compared visual acuity measured with letters versus with optotypes such as Es had used steadily-presented targets rather than briefly-presented masked targets.

Procedure

Trials were initiated by the subject using a bar-press. The 19 subjects were practiced for 12 to 15 minutes before experimental data were collected for 50 to 65 minutes. They were instructed to report the direction of the offset E by pressing the right or left pointing arrow on the keyboard. They were informed that only one target (E or backward E) would appear on each trial, that there was always a target, and that they should make their best guess if they were uncertain. As in

Experiment 1, subjects were run in a blocked design, in which they monitored all four locations or just two of them (top-bottom or left-right). The duration of the offset Es, as depicted in Figure 9.1, was 50 ms.

Results

Latencies were *faster* in monitor 2 than in monitor 4 ($F_{1,19} = 9.2, p = 0.007$) at all locations. The benefit of focusing was larger in top-bottom than in left-right locations ($F_{3,57} = 2.71, p = 0.049$), as shown in Figure 9.6. Since the focusing effect in the left-right locations was within the standard error bars, we conclude that focusing speeds E detection at top and bottom. This pattern confirmed the pilot experiment in which the advantage for focusing top and bottom held across stimulus durations randomized over trials from 50ms to 75, 100, 125, to 150 ms (Fine & Reeves, 2001).

Accuracy was high (average 93 percent) and did not vary with number monitored, location, or their interaction when arc-sine transformed (all *F*'s with $p >$ 0.15), and so are not graphed. There was a small Simon effect, in that the forward E was slightly more accurate on the right and the backward E on the left, but this did not affect the overall conclusion: focusing attention *improved* the detection of masked offset Es in the top-bottom locations, which is in the direction opposite to that for identifying letters. We remark here that the pilot experiment on 15 subjects had revealed that accuracy rose from 92 percent to 94 percent as stimulus duration was increased from 50, 75, 100, 125, to 150 ms (Fine & Reeves, 2001). Such high accuracy in the same conditions is consistent with the Es being detected at an earlier level of processing than the letters in Experiment 1.

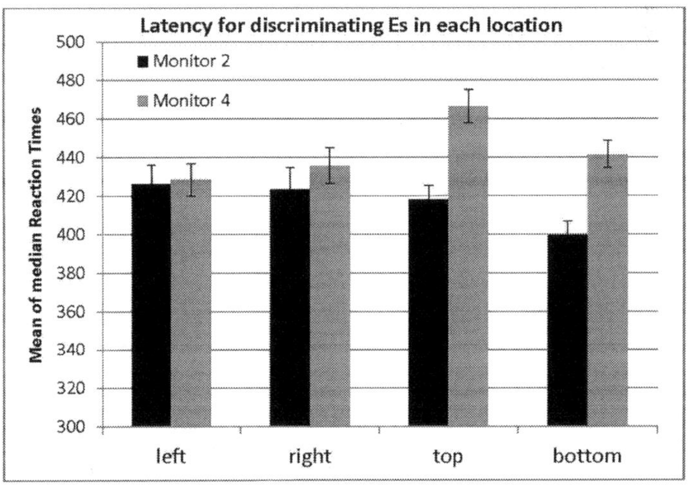

FIGURE 9.6

Latency for reporting the orientation of offset Es in the diagonal format, when monitoring two or all four positions, in Experiment 3. Latency was faster when monitoring 2 than 4 locations. Bars give ±1 standard error. (Accuracy was uniformly high and is not plotted).

Experiment 4: Consonant/Vowel Discrimination

That the letters and optotypes (Es) gave opposite results might be due either to the nature of the stimuli or to the task. Experiment 4 was designed to isolate the effect of the task by obtaining results with letters when the overt task, like that for the Es, was binary. Ten letters were used, divided into vowels (A E I O U) and consonants (H B T C V). Eighteen subjects reported on each trial whether the offset letter had been a vowel or a consonant (CV discrimination). 18 more subjects identified the target letter, as in Experiment 1. The diagonal format of Experiment 2 was used, with stimuli at 5 deg. eccentricity. Letters were presented for 50 ms and followed by placeholders for 250 ms.

Results

Mean consonant/vowel (CV) discrimination accuracy was 69 percent, considerably worse than accuracy for identifying one of the ten letters (81 percent), presumably because CV discrimination requires identifying a letter before deciding on its category. Critically, monitor 4 remained better than monitor 2 for CV discrimination, by 5.2 percent at lower left, 4.4 percent at upper left, 4.8 percent at upper right, and 1.9 percent at lower right (not plotted to save space); however, the usual number monitored by location interaction was not significant ($p = 0.11$). Once again, focusing was deleterious to accuracy, just as in Experiments 1 and 2. We conclude provisionally that it is not the task, i.e., a binary or n-ary choice, but rather the level of processing of the stimulus (E versus letter) that determines if attentional focusing will help or hinder the subject in these experiments.

Model

Focusing helps the discrimination of simple offset stimuli (Es) but hurts the identification of more complex offset stimuli (letters). How can attention have opposite effects of this sort? Our signal-detection-based explanation is simply that focusing attention always improves processing, but differentially for the signal and for the noise.

Before proceeding, we address four background issues. First, we assume that processing of our brief displays is 'effectively' parallel across the visual field, following an enormous literature on parallel processing (Gardner, 1973; Nakayama & Bravo, 1992, even used the term 'massively parallel'). Why 'effectively'? Serial movements of attention can aid focusing (e.g., Najemnik & Geisler, 2008), but if identification or detection were serial but too slow to visit more than two or three locations before stimulus termination, then accuracy for reporting from two positions would always exceed accuracy for reporting from four locations, *contra* our data. A *rapid* serial attentive process (e.g., 12 ms per item: Liss & Reeves, 1983), cannot be rejected in this way, as it could visit all four locations within the duration of the stimulus, so we cover both cases with the term 'effectively parallel', these being equivalent for our model. Second, we assume 'independent channels' (Gardner, 1973), in that our 4 locations, being separated by 7 deg., were far enough apart to act as independent sources of information, as shown psychophysically and even electro-physiologically (Sutter, 2001). Third, the guessing rate (50 percent for the Es and CVCs, 10 percent or 5 percent for identifying 1 of 10 or 20 letters) was always independent of number monitored. Therefore, any effect of monitoring on accuracy (hence d', given the forced-choice method) can only be due to the focusing or diffusion of attention, not to variations in the guessing rate. Fourth, our results with letters cannot be explained by attention to the wrong locations or wrong spatial frequency band (Yeshurun & Carrasco, 1998), as the target stimuli and placeholders were all in the same band and had the same locations and eccentricities.

With these preliminaries in hand, we define S/N as the signal to noise ratio, letting S stand for signal strength, α for the level of attention, and R for the response. The total noise, N, is given by $N = Ne + Ni$, where $Ne \geq 0$ is the external noise due to placeholders and $Ni > 0$ is the internal noise from random neural activity. The model is

$$R = \alpha\, S/\sqrt{[(\alpha^2\, Ne)^2 + Ni^2]} \qquad \text{Eq. 1}$$

R is assumed proportional to d', so $R = 0$ when $S = 0$, and R is positive but finite when $S > 0$ as $Ni > 0$. Ne and Ni are necessarily independent, so their variances add; the square root returns the standard deviation in the same units as S. The mean and variance of the noise are assumed to be in fixed ratio, as is true for visual detections over a wide range of luminances (Reeves et al., 1997), so need not be modelled separately.

We also assume that RTs are inverse with response strength, as in the LATER model of Carpenter, Reddi, and Anderson (2009), so that

$$RT = b/R + RTo \qquad \text{Eq. 2}$$

where b is the constant of proportionality that scales 1/R to latency and RTo is a fixed motor component. Equations 1 and 2 imply that an increase in the S/N ratio improves performance by speeding the response and increasing sensitivity in tandem.

Inspection of Equation 1 shows that when internal noise dominates ($Ne << Ni$), an increase in α increases S more than N and therefore increases R, so attention helps, whereas when external noise dominates ($Ne >> Ni$), an increase in α increases N more than S, so attention hurts. In the case of a balance such that $Ni = Ne\,(\alpha^2/2)$, neither speed nor accuracy will change with attention, so α has no effect. Note that explaining how attention can hurt or help performance, or have no effect, requires a model in which attention affects S differentially from N, in contrast to models in which attention always helps (e.g., Palmer, 1995).

To fit Equations 1 and 2 to data, we equate increasing attention in the model with focusing by the subject: specifically, α doubles in focusing, so α = 1.0 in monitor 2 and α = 0.5 in monitor 4. We can then construct various R versus S curves. We assume constant internal noise, such that Ni=1.0. When Ne = 0.1, so Ne << Ni, the noise is mostly internal, and the two dotted theoretical curves in Figure 9.7 show that attending to 2 locations (small grey diamonds) is *superior* to attending to all 4 (small grey triangles), at all levels of S (horizontal axis). When Ne =10, the noise is mostly external and the solid theoretical curves in Figure 9.7 show that attending to 2 positions (small black diamonds) is *inferior* to attending to 4 positions (small black triangles).

Large unconnected symbols on the right side of Figure 9.7 plot the four corresponding response *rates*, that is, the measured mean RTs in seconds from Experiments 1 and 3 converted to rates by Equation 2 with RTo = 0.4 sec and b = 0.17. These values for RTo and b have no independent justification (though RTo seems plausible) other than that they scale the rates to the theoretical plots, rate being chosen over RT to conform to the plotting convention that upwards is better. The rate of processing letters (large black symbols) is faster with diffuse attention (Mon 4)

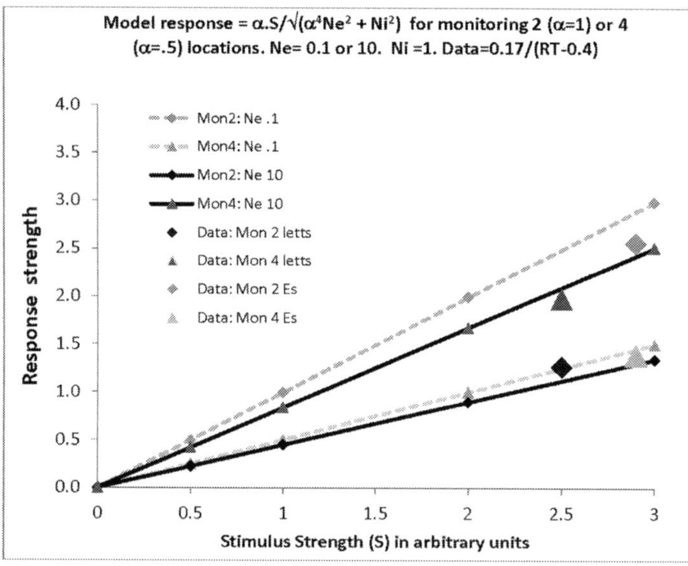

FIGURE 9.7

Theoretical response strength, R, is plotted against signal strength, S, from Equation 1, for four conditions: monitor (Mon) 2 (small diamonds) or Mon 4 (small triangles) with low external noise (Ne = 0.1) (grey symbols), when focusing helps, or high external noise (Ne = 10) (black symbols), when focusing hinders. Large diamonds (RTs in Mon 2) and triangles (RTs in Mon 4) on the right plot latencies converted to rates according to Equation 2 with b = 0.17 and RTo = 0.40 sec; black symbols for letters, grey symbols for Es.

than focused attention (Mon 2), whereas for Es (large grey symbols), the reverse is true. The model provides a phenomenological account for why attention can both help and hinder processing, and roughly approximates the four data points, but obviously more parametric data will be needed to refine and perhaps elaborate the model in the future. In particular, the model assumes that latency increases with detectability, as expressed in Equation 2, a simplification which ignores the type of trade-off found in Experiment 2.

Conclusions

Focusing attention in the visual periphery can slow RTs and reduce accuracy for identifying offset letters but can help detection of offset optotypes (forward and backward Es). Therefore simply inferring increased attention from improved performance can be erroneous, which is our take-home message. Given that our subjects were appropriately motivated and instructed, we relied on signal detection modeling to understand the results.[1] We supposed that selective attention boosts both the signal and the external noise, but does so unequally: external noise (due to the placeholders) is boosted by more than the signal when letters are identified, hurting performance, but by less than the signal when Es are detected, aiding performance. This applied to every position tested except directly left and right of fixation, when focusing had no effect, perhaps due to reading habits; in the model, the lack of a focusing effect is explained if focusing boosts both signal and noise equally.

The model is purely formal and does not address mechanism; for example, the type of masking and the level of processing are not explicitly addressed. However, any possible mechanism should be compatible with the model, if the model holds up. As a possible mechanism, we speculate that extra attention must be paid to the placeholders to identify the offset letters, given that the placeholders act as patterned masks (Turvey, 1973; Haber, 1970). Endogenous attention, needed to process offset letters, appears to act as a gate with a minimum opening/closing time of approximately 160 ms (Reeves & Sperling, 1986). Thus focused attention may grab the after-coming masks as well as the targets and thereby boost the external noise higher than the signal when the signal requires more time to process – as with identifying letters compared to detecting Es. This notion is compatible with varying Ne and fixing Ni, as illustrated in Fig. 7. An alternative, also compatible with the model, assumes that external noise (Ne) is fixed by the placeholders, but that internal noise (Ni) is greater when detecting E's than identifying letters.

Will our results generalize? We suspect, but do not know, that other situations in which the observer must extract a known endogenous target from a background of noisy patterns will, as in our experiments, reveal cases in which focusing reduces efficiency.

Acknowledgments

The data were collected while the first author was at the Schepens Eye Research Institute. The research was supported, in part, by donations from the William Randolph Hearst and the Joseph E. and Marjorie P. Jones Foundations to the first author. Thanks to Cathleen Moore for helpful discussions of these data.

Note

1 Scharf et al. (1997) measured thresholds for detecting pure tonal signals in noise both before and after the OCB (olivochochlear bundle) was cut surgically in Menier's syndrome patients. Pre-operative thresholds were better for expected than unexpected tones by about 8dB, as if attention aids hearing. However, post-operative thresholds were the same as pre-operative ones for expected tones, improving only for unexpected tones. Since OCB feedback mediates the effects of attention in the auditory system, the authors concluded that auditory attention in normals suppresses unattended signals but does not enhance attended ones. No similar operation exists in vision, and so our model can only predict the S/N ratio, not the signal or noise level *per se*.

References

Bahcall, D. O., & Kowler E. (1999). Attentional interference at small spatial separations. *Vision Research, 3*, 7–86.

Blaser, E., Pylyshyn, Z. W., & Holcombe, A. (2000). Tracking an object through feature-space. *Nature, 408*, 196–199.

Carpenter, R. H. S., Reddi, B. A. J., & Anderson, A. J. (2009). A simple two-stage model predicts response time distributions. *Journal of Physiology, 587*, 4051–4062.

Carrasco, M., Williams, P. E., & Yeshurun, Y. (2002). Covert attention increases spatial resolution with or without masks: Support for signal enhancement. *Journal of Vision*, 2, 467–479.

Dinsmoor, J. A. (1985). The role of observing and attention in establishing stimulus control. *Journal of the Experimental Analysis of Behavior, 43*, 365–381.

Donaldson, M. J., & Yamamoto, N. (2016). Detection of object onsets and offsets: Does the primacy of onset persist even with bias for detecting offset? *Attention, Perception & Psychophysics, 78*, 1901–1915.

Eckstein, M. P. (1998). The lower visual search efficiency for conjunctions is due to noise and not to serial attentional processing. *Psychological Science, 9*, 111–118.

Egly, R., Driver, J., & Rafal, R. D. (1994). Shifting visual attention between objects and locations: Evidence from normal and parietal lesion subjects. *Journal of Experimental Psychology: General, 123*, 161–177.

Eriksen, C. W., & Yeh, Y. Y. (1985). Allocation of attention in the visual field. *Journal of Experimental Psychology: Human Perception and Performance, 11*, 583–597.

Fine, E., & Reeves, A. (2001). Monitoring more locations increases reaction time, but only above and below fixation. *Investigative Ophthalmology and Visual Science, ARVO Suppl.* 42.

Fine, E., & Reeves, A. (2002). Processing benefits from diffuse attention when the stimuli are harder to discriminate. *Vision Sciences Society*, Annual Meeting, Sarasota, FL.

Gardner, G.T. (1973). Evidence for independent parallel channels in tachistoscopic perception. *Cognitive Psychology, 4*, 130–155.
Greenberg, G. Z., & Larkin, W. D. (1968). Frequency-response characteristics of auditory observers detecting signals of a single frequency in noise: The probe-signal method. *Journal of the Acoustical Society of America, 44*, 1513–1523.
Grindley, G. C., & Townsend, V. (1968). Voluntary attention in peripheral vision and its effects on acuity and differential thresholds. *Quarterly Journal of Experimental Psychology, 20*, 11–19.
Haber, R. N. (1970). Note on how to choose a backward mask. *Psychological Bulletin, 74*, 373–376.
Hughes, H. C., & Zimba, L. D. J. (1985). Spatial maps of directed visual attention. *Journal of Experimental Psychology: Human Perception and Performance, 11*, 409–430.
James, W. (1890/1950). *The Principles of Psychology*, G. Miller (Ed.). New York: Dover Publications.
Jonides, J., & Yantis, S. (1988). Uniqueness of abrupt visual onset in capturing attention. *Perception & Psychophysics, 43*, 346–354.
Kanwisher, N., & Wojciulik, E. (2000). Visual attention: Insights from brain imaging. *Nature Reviews Neuroscience, 1*, 91–100.
Klein, R., & McCormick, P. (1989). Covert visual orienting: Hemifield-activation can be mimicked by zoom lens and midlocation placement strategies. *Acta Psychologica, 70*, 235–250.
Kurylo, D., Reeves A., & Scharf, B. (1996). Expectancy of line segment orientation. *Spatial Vision, 10*, 149–162.
Liss, P., & Reeves, A. (1983). Interruption of dot processing by a backward mask. *Perception, 12*, 513–529.
Lu, Z. L., & Dosher, B. (1999). External noise distinguishes attention mechanisms. *Vision Research, 38*, 1183–1198.
Lu, Z. L., & Dosher, B. (2008). Characterizing observers using external noise and observer models: Assessing internal representations with external noise. *Psychological Review, 115*, 44–82.
Marcel, A. J. (1983). Conscious and unconscious perception: Experiments on visual masking and word recognition. *Cognitive Psychology, 15*, 197–237.
Najemnik, J., & Geisler, W. S. (2008). Eye movement statistics in humans are consistent with an optimal search strategy. *Journal of Vision, 8*, 1–14.
Nakayama, K., & Bravo, M. (1992). The role of attention in different visual search tasks. *Perception & Psychophysics, 51*, 465–472.
Owens, C., & Spehar, B. (2016). Do sudden onsets need to be perceived as new objects to capture attention? The interplay between sensory transients and display configuration. *Attention, Perception, & Psychophysics, 78*, 1916–1925.
Palmer, J. (1995). Attention in visual search: Distinguishing four causes of set-size effects. *Current Directions in Psychological Science, 4*, 118–123.
Posner M. I., Nissen M. J., & Ogden, W. C. (1978). Attended and unattended processing modes: The role of set for spatial location. In H. L. Pick & E. J. Saltzman (Eds), *Modes of Perceiving and Processing Information* (pp. 137–157). Hillsdale, NJ: Erlbaum.
Reed, A. (1973). Speed-accuracy trade-off in recognition memory. Science, *181*, 574–576.
Reeves, A., & Scharf, B. (2010). Auditory frequency focusing is very rapid. *Journal of the Acoustical Society of America, 128*, 795–803.
Reeves, A., & Sperling, G. (1986). Attentional theory of order information in short-term visual memory. *Psychological Review, 93*, 180–206.

Reeves A., Wu S. & Schirillo J. (1997). The effect of photon noise on the detection of white flashes. *Vision Research, 38*, 2639–2647.

Santhi, N., & Reeves, A. (2004). The roles of distractor coherence and target certainty on visual search: A signal detection model. *Vision Research, 44*, 1235–1256.

Scharf, B., Chays, A., & Magnan, J. (1997). The role of the olivocochlear bundle in hearing: Sixteen case studies. *Hearing Research, 103*, 101–122.

Scharf, B., Reeves, A., & Suciu, J. (2007). The time required to focus on a cued signal frequency. *Journal of the Acoustical Society of America, 121*, 2149–2157.

Skelton, J. M., & Eriksen, C. W. (1976). Spatial characteristics of selective attention in letter matching. *Bulletin of the Psychonomic Society*, 7, 136–138.

Sutter, E. E. (2001). Imaging visual function with the multifocal m-sequence technique. *Vision Research, 41*, 1241–1255.

Turvey, M. (1973). On peripheral and central processes in vision: Inferences from an information-processing analysis of masking with patterned stimuli. *Psychological Review, 80*, 1–52.

Yeshurun, Y., & Carrasco, M. (1998). Attention improves or impairs visual performance by enhancing spatial resolution. *Nature*, 396, 72–75.

10

HIGH-LEVEL INFLUENCES ON PERCEPTION

Then and Now

Mary A. Peterson
University of Arizona

I. Introduction

Naomi Weisstein's work was groundbreaking. In the 1970s, Weisstein and her colleagues showed that psychophysical techniques could be used to investigate high-level as well as low-level processes involved in visual perception (see Weisstein & Maguire, 1978 for review). Their findings showed that, contrary to the well-entrenched serial, hierarchical, view of perception, high-level processing "filters, overrides, or enhances the perception of the details in an image" (Weisstein & Maguire, 1978, p. 243). Research supporting this view has accumulated since Weisstein's pioneering work.

In Section II of this chapter, I review some of Weisstein's seminal work on high-level effects on the perception. In Section III, I discuss recent research revealing effects of past experience and global context on figure assignment. In Section IV, I review Weisstein's own work on figure assignment and interpret it in a Bayesian framework in which learning produces priors for figure assignment. Section V summarizes the chapter and concludes that Weisstein's influence is enduring.

II. Weisstein's Seminal Work on High-Level Effects on Perception

High-level effects on visual perception can be shown in a variety of ways. One way is via demonstrations that global configuration or context affects how local elements are perceived. The reasoning behind this approach is as follows: low-level units respond to only a small portion of the visual input, whereas units higher in in the visual hierarchy respond to larger, more global, portions of the input. Therefore, an effect of global configuration on local perception arises from higher levels of processing. In one of the earliest experiments demonstrating global influences on the perception of local detail, Weisstein and Harris (1974) showed observers very brief exposures of one of four diagonal target lines and asked them to report the line orientation (a local detail). The diagonal lines were presented along with context lines that were identical for all orientations of the test lines; hence, as low-level stimuli, the context lines were not informative regarding the orientation of the target line. Typically, the addition of other lines to the test display interferes with observers' ability to report which target line was presented. However, Weisstein and Harris found that one particular set of context lines enhanced line orientation discrimination by as much 19 percent (see Figure 10.1). The beneficial context lines formed overlapping squares that, when combined with the four possible diagonal target lines formed four different three-dimensional configurations. The higher-level global organization of the lines into three-dimensional configurations enhanced line orientation accuracy. This *object superiority effect* was a powerful demonstration that global configuration affects the perception of local details (Weisstein & Harris, 1974; see also the *word superiority effect*: Reicher, 1969; Wheeler, 1970).

Weisstein and her colleagues also demonstrated global influences on the perception of local details using illusory contours (Weisstein et al., 1974), amodal completion (Weisstein, 1970, 1973), and phantom motion aftereffects (Weisstein et al., 1977; cf., Tynan & Sekuler, 1975). As a unifying account of these effects,

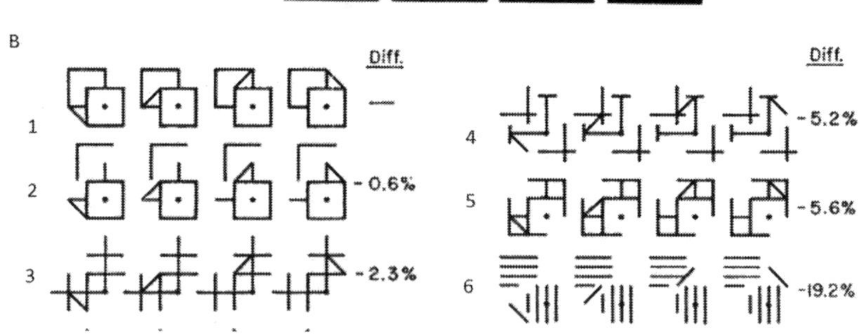

FIGURE 10.1 Adapted from Weisstein and Harris (1974), Figure 2.

(A) The four target lines (a–d); the context lines (e), and the configurations formed by combining the four target lies with the context lines (f–i). (B) Sample configurations formed by combining the four target lines with various context lines. The values in the column labeled "Diff" indicate the extent to which accurate line detection decreased in conditions 2–6 compared to condition 1.

Weisstein and Maguire (1978) hypothesized that, "Higher level processes may be feeding back to lower-level mechanisms, and our results may reflect the outcome of successive computations by these mechanisms" (p. 252). At that time, the idea that feedback plays a role in visual perception was not yet broadly acknowledged; Felleman and VanEssen's (1991) anatomical data were yet to revolutionize theorizing. Accordingly, Weisstein and Maguire cautiously acknowledged that their results didn't demand an explanation in terms of feedback. The careful psychophysical work of Naomi Weisstein and her colleagues paved the way for other behavioral experiments showing high-level influences on perception, including my own work on figure assignment.

III. Recent Evidence of High-Level Influences on Figure Assignment/Object Detection

Many scenes comprise multiple borders between regions. A border shared by two regions is often perceived as the bounding contour of only one of them. The region to which the border is assigned is perceived as a shaped entity, called a *figure* in classic terms, whereas the other region, lacking a bounding contour appears locally shapeless near the border and seems to simply continue behind the figure as its background. The traditional view is that figure assignment is determined early in the course of perception based on local, low-level, image properties. In this section, I briefly review recent work from my laboratory revealing that

(1) memories for previously seen object configurations are priors for figure assignment; and
(2) figure assignment is affected by scene-wide information.

Both lines of evidence are consistent with the view that high-level processing "filters, overrides, or enhances the perception of the details in an image" (Weisstein & Maguire, 1978, p. 243).

Scene-Wide Influences on the Figure Assignment

The Gestalt psychologists had demonstrated that image-based properties such as convexity and symmetry were figural "cues" in that regions with these properties were much more likely to be perceived as the objects in the scene—the figures—than adjacent regions that were concave and asymmetric, respectively. They proposed that infants were innately endowed to respond to these, and other, cues. The displays the Gestalt psychologists used to demonstrate the influence of these image-based features had many (eight) alternating black and white regions like those in Figure 10.2, top row, right column.

Although the Gestalt psychologists are known for their theoretical position that the whole is different from the sum of its parts and many demonstrations to this effect, they did not investigate whether effects of image properties such as

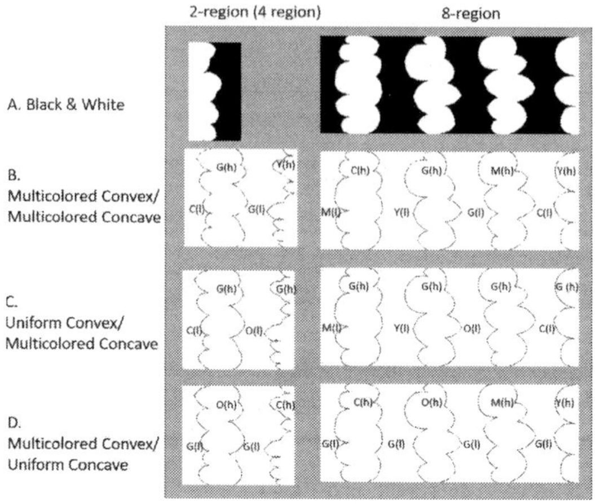

FIGURE 10.2

(A) Black and white displays. *A sample two-region display is shown on the left, and a sample eight-region display is shown on the right. Displays were shown one at a time on a medium gray backdrop. Here, the convex regions are white, and the concave regions are black. The contrast polarity of the convex and concave regions was balanced across a set of 64 displays. (B–D) Schematic depictions of the colored displays used by Peterson and Salvagio (2008).*[1] *For colored displays, four-region displays were the smallest region number condition that could be tested. Four-region displays are shown on the left; eight –region displays on the right. The capital letters denote the color of the fill: C = cyan, G = gray, M = magenta, Y =yellow, O = orange. Note that the concave and convex regions always differed in luminance, one was high luminance (h); the other was low luminance (l). The high versus low-luminance of the convex and concave regions was balanced across a set of 72 displays. (B) A schematic depiction of the* Multicolored Convex/Multicolored Concave displays *where no two regions of a given type were filled with the same color. (C) A schematic depiction of the* Uniform Convex/Multicolored Concave displays *where all convex regions were uniformly filled (here, in gray), and the concave regions were multicolored. (D) A schematic depiction of the* Multicolored Convex/ Uniform Concave displays *where the convex regions were multicolored, and the concave regions were uniformly filled (here, in gray).*

convexity or symmetry were influenced by the global context in which they were tested. Peterson and Salvagio (2008) and Mojica and Peterson (2014) showed that global context exerts a large effect. For instance, Peterson and Salvagio replicated experiments showing that the convex regions are perceived as figures on approximately 90 percent of trials (e.g., Kanizsa & Gerbino, 1976) when they presented displays with eight alternating black and white (B/W) convex and concave regions like those in Figure 10.2A on a medium gray background.[2] Peterson and Salvagio also investigated figure reports made by observers viewing displays with six, four and two alternating B/W convex and concave regions. (Different observers viewed displays in the different region-number conditions.) They found that, as the number of alternating convex and concave regions in the B/W displays reduced systematically from eight to two, so too did observers' reports that they

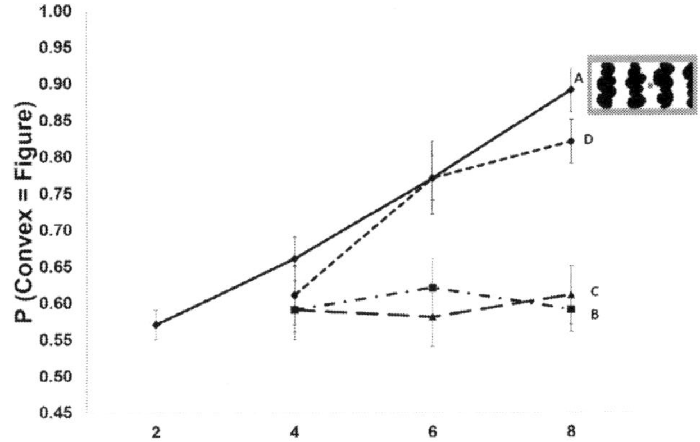

FIGURE 10.3 The probability of convex figure reports as a function of the number of regions in the displays.

(A) Black and white displays. (B) Multicolored Convex/Multicolored Concave displays. (C) Uniform Convex/Multicolored Concave displays. (D) Multicolored Convex/ Uniform Concave displays.

perceived the figure on the convex side of the borders (see Figure 10.3, line A). For two-region displays like the one on the left in the top row of Figure 10.2A, the figure was perceived on the convex side of the central border on only an average of 57 percent of trials. These region-number effects showed that something about the scene-wide context of the eight-region B/W displays amplifies the effectiveness of the image property of convexity as a figural cue (see also Lass et al., 2017).

In B/W displays, both convex and concave regions are uniformly filled—all regions of a given type are either black or white. Peterson and Salvagio (2008) tested whether uniform fill was required in one or both types of regions by comparing performance with:

(A) B/W displays.

(B) Multicolored convex/multicolored concave displays in which there was no uniformity of fill in either the convex or the concave regions. (See Figure 10.2B for a schematic version and Peterson & Salvagio, 2008, Figure 4, top row, for a colored example of these stimuli.) If region-number effects are due to probability summation alone and uniform fill is unnecessary, then region-number effects like those observed for B/W displays should be observed with these MC-convex/MC-concave displays. As can be seen in Figure 10.3, line B, region number effects were not obtained when observers viewed multicolored-convex/multicolored-concave displays, ruling out probability summation as the cause.

(C) Uniformly filled convex/multicolored concave displays in which all convex regions were the same color whereas concave regions were multicolored. (See Figure 10.2C for a schematic version and Peterson & Salvagio, 2008, Figure 4, middle row, for a colored example of these stimuli.) If region number effects occur because convex regions group together by virtue of similar color, then region-number effects like those observed for B/W displays should be observed with these uniformly filled convex/multicolored concave displays. As can be seen in Figure 10.3, line C, region number effects were not obtained when observers viewed uniformly filled convex/multicolored concave displays, ruling out grouping of the convex regions as the cause.

(D) Multicolored convex/uniformly filled concave displays in which all concave regions were the same color whereas convex regions were multicolored. (See Figure 10.2D for a schematic version and Peterson & Salvagio, 2008, Figure 4, bottom row, for a colored example of these stimuli.) If region number effects require some form of grouping of concave regions, then region-number effects like those observed for B/W displays should be observed with these multicolored convex/uniformly filled concave displays. As can be seen in Figure 10.3, line D, region number effects were obtained when observers viewed multicolored convex/uniformly filled concave displays. Thus, uniform fill in the concave regions was required for the emergence of region-number effects; the uniformity versus heterogeneity of the fill of the convex regions was irrelevant.[3]

Based on a Bayesian observer analysis of these results, Goldreich and Peterson (2012) proposed that region-number effects were due to the joint operation of two "priors":

(1) A convexity prior such that objects are more likely to be perceived on the convex rather than the concave side of a border. Goldreich and Peterson's analysis indicated an odds-ratio of 1.77:1 for convex over concave figures, which was substantially lower than expected based on the Gestalt demonstrations. Burge, Fowlkes, and Banks (2010) provided a strikingly similar estimate from an analysis of scenes in the real world.

(2) A background prior such that disconnected regions that are uniformly filled rather than heterogeneously filled are likely to be perceived as portions of a background surface, and increasingly so as they increase in number. Based on Peterson and Salvagio's research, Goldreich and Peterson assigned the uniformly filled background prior an odds ratio of 1.44:1. To our knowledge, no other estimates of a uniform fill/background odds ratio exist, but others have observed that color varies less across the surface of a single object than it does between objects in natural scenes (Fine et al., 2003; Ing et al., 2010).

Yin, Kellman, and Shipley (1997, 2000) showed that uniformity of color enhances amodal completion (cf., Su et al., 2010). It is likely that amodal completion is necessary in order for disconnected uniformly filled regions like those in the displays used by Peterson and colleagues to be perceived as a background behind the convex regions. Mojica and Peterson (2014) obtained further evidence that amodal completion plays a role in region-number effects when they investigated whether region-number effects were obtained for the figural property of symmetry. They found that the shape of the bounding contour of the entire display was critical; region number effects were obtained only when the outer border of the display was consistent with amodal completion of the asymmetric regions into a surface behind the symmetric regions. Burrola and Peterson (2014) extended this finding to displays with alternating convex and concave regions (see Figure 10.4).

The evidence for global influences on figure assignment due to the number, type, and fill of alternating regions as well as the shape of the outer display border is consistent with Weisstein's evidence that global configurations affect the *detection* of line orientation (e.g., Weisstein & Harris, 1974). In my view, figure-assignment is essentially object detection, inasmuch as it entails detecting where an object lies with respect to a border; no special figure-ground segregation process exists (cf., Peterson, 2003). Our work demonstrates that various priors interact to determine where an object lies with respect to a border; some priors are supported by global display-wide features, others by local features. Weisstein and her colleagues were among the first to demonstrate that the global reach of amodal completion implied high-level influences on perception (Weisstein, 1970, 1973; Weisstein et al., 1977, 1972). It has recently been shown that amodal completion involves

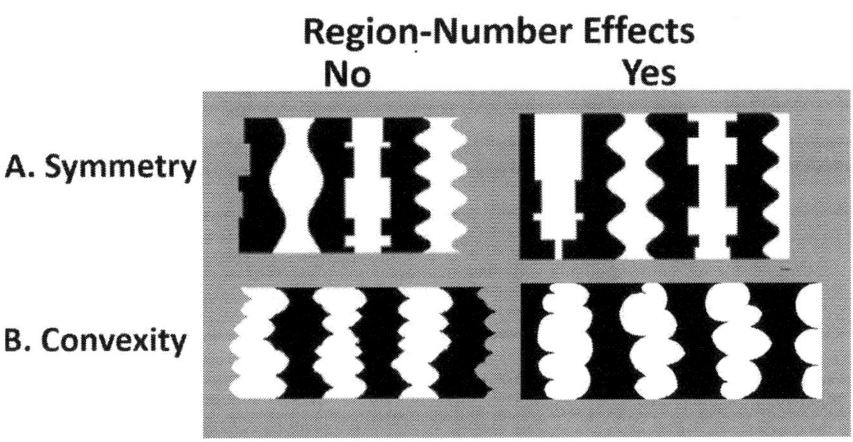

FIGURE 10.4 The rectilinear borders of the displays on the right afford amodal completion of the region truncated at the border, and hence, of the asymmetric or concave regions in (A) and (B) whereas the articulated borders of the displays on the left do not.

feedback from higher to lower levels of visual processing (e.g., Peterson & Salvagio, 2017; Wyatte et al., 2012).

In the remainder of this section, I will review evidence that memory for familiar configurations serves as an object detection prior as well.

Object Memory Influences on Figure Assignment

In the early 1990s, my students and I published the first evidence that memories of previously seen objects serve as priors for object detection (aka "figure assignment;" Gibson & Peterson, 1994; Peterson & Gibson, 1991, 1993, 1994a, 1994b; Peterson et al., 1991). For instance, for displays like Figure 10.5A comprising two equal-area regions separated by an articulated border, participants were substantially and significantly more likely to detect an object on the side of the border where a portion of a well-known mono-oriented real-world object was suggested when they viewed stimuli with the well-known object in its typical upright orientation rather in an unfamiliar upside down orientation (see Figure 10.5C). The orientation manipulation held constant the features and parts of the potential object, while it varied familiarity because, by definition, participants are more familiar with mono-oriented objects in their typical upright orientation.

To investigate whether these were local effects, due to the familiarity of the parts or whether they were more global effects of configured collections of familiar parts, Peterson et al. (1991; Gibson & Peterson, 1994) compared object detection performance when observers viewed intact upright displays like

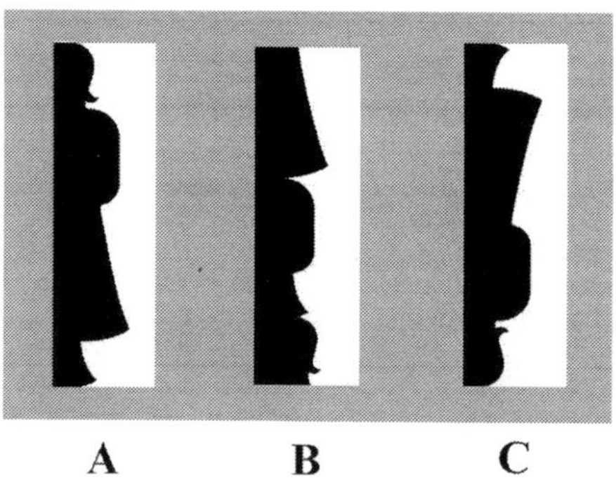

FIGURE 10.5

(A) A display divided into two equal area regions by a central border. One region (here, black, left) depicts a portion of an upright well-known (familiar) object (here, a woman). (B) The black region on the left depicts a portion of a "part-rearranged" novel object created by dividing the familiar object in (A) into parts at concave cusps and spatially rearranging those parts into a novel configuration. (C) The black region on the left depicts an upside down version of A. These displays were shown on a medium gray backdrop. The black/white polarity and left/right location of the critical regions discussed above were balanced in the set of test displays.

those in Figure 10.5A versus control displays created by spatially rearranging the parts of the well-known objects into novel configurations (part-rearranged novel objects) as shown in Figure 10.5B. Participants were substantially and significantly more likely to detect the object on the critical side of the border when the same familiar parts were arranged in the familiar configuration rather than a novel configuration. These "familiar configuration" effects revealed that memories of configured objects are accessed before object detection, and constitute priors for where objects lie with respect to borders in a scene. Similar effects have been obtained for observers as young as 5 years of age on average (Spanò et al., 2016) and as old as 71 years of age on average (Peterson et al., 2000, 1998). Peterson and Gibson (1994b) showed that object memories are not necessarily dominant in object detection. Indeed, Peterson and Gibson (1994b) found that the weight assigned to the priors of familiar configuration and symmetry were approximately equal. Accordingly, Peterson (1994) proposed that object memories constitute one of many priors for where objects lie with respect to a border.

The studies just reviewed showed that high-level influences on perception arise from memories accessed in the course of object detection as well as from global scene-wide interactions. It is not yet completely clear where in the visual hierarchy the relevant object memories are represented; more research is needed to answer this question. Peterson (2003) speculated the relevant brain region might be V4. However, recent neuropsychological and neuroimaging work implicates even higher-level brain regions in familiar configuration effects on object detection. Specifically, Barense et al. (2012) reported that patients with damage to the perirhinal cortex (PRC) of the medial temporal lobe (MTL) showed effects of familiar *parts* on figure assignment: their patients reported perceiving the region depicting a part-rearranged novel object as figure (sample shown in black on the left in Figure 10.5B) as often as they reported perceiving the region depicting a familiar configuration as figure (sample shown in black on the left Figure 10.5A). Age-matched control participants did not show this pattern; they, like the participants originally tested by Peterson et al. (1991) and Gibson and Peterson (1994) reported perceiving the region depicting a familiar configuration as figure substantially and significantly more often than the region depicting a part-rearranged novel configuration. Barense et al. hypothesized that the intact PRC responds to the familiarity of collections of parts as well as to spatial configurations, and that feedback from the PRC modulates familiar part responses in the early visual cortex (EVC), enhancing the response when familiar parts are in a familiar configuration versus a novel configuration. To test these hypotheses, Cacciamani, Wager, Peterson, and Scalf (2017) and Peterson, Cacciamani, Barense, and Scalf (2012) measured fMRI while participants viewed displays where the critical regions were biased to be perceived as figure by contrast and small area; their task was to report whether the displays depicted well-known objects or novel objects. The results supported Barense et al.'s predictions, showing that:

(1) the PRC response varies as a function of the familiarity of the configuration and of the collection of parts constituting the configuration, with the highest response when familiar parts were arranged into a familiar configuration and the lowest response when familiar parts were arranged into a novel configuration, with an intermediate response to control novel stimuli (see Figure 10.6); and
(2) that the EVC response for the same familiar parts was higher when they were arranged in a familiar configuration rather than a novel configuration. Results of fMRI experiments are only suggestive regarding feedback although, consistent with the feedback hypothesis, a functional connectivity analysis performed by Cacciamani et al. (2017) on their data revealed that left hemisphere PRC was functionally connected to left hemisphere EVC (specifically V2). We await direct evidence that feedback plays a role in integrating the familiarity of parts and larger configurations.

The results from my laboratory were not explicitly predicted by Weisstein's earlier research, yet they are consistent with her work in that they show that context and high-level object memories affect object detection (aka figure assignment), which was long held to occur at a low level in the visual hierarchy. Moreover, the recent results implicating the PRC in the integration of responses to parts and wholes implies that a brain structure at the top of the ventral processing pathway is involved in object memory influences on object detection.

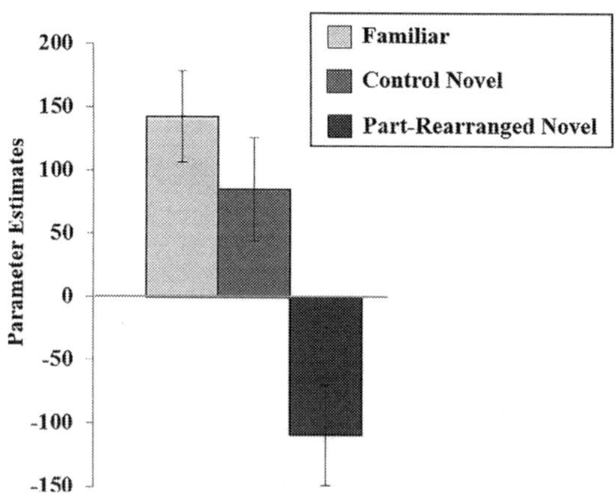

FIGURE 10.6 Parameter estimates from Cacciamani et al. (2017) when observers viewed portions of familiar objects (light gray), part rearranged novel objects (dark gray), and control novel objects (medium gray) which were inverted versions of the part rearranged novel objects.

IV. Weisstein's research on figure-ground perception

Weisstein and her colleagues (e.g., Brown &Weisstein, 1988; Klymenko et al., 1989; Wong & Weisstein, 1982; 1983) conducted a series of experiments investigating the relationship between figure assignment and spatial frequency. These experiments can be understood today as evidence that observers learn that different types of information are typically extracted from figures (i.e., objects) and grounds; this past experience

(1) prioritizes differential types of information on figures and grounds; and
(2) operates as a figural prior.

When they were published, these experiments were not interpreted as evidence that past experience influences figure assignment, however.

Wong and Weisstein (1982, 1983) showed that different types of information are prioritized for figures versus grounds in experiments using a version of the Rubin vase-faces stimulus. They instructed observers to maintain fixation on a point on the center of the vase, and to press a key when they achieved an instructed interpretation, which varied across trials (i.e., to see either the center region or the surrounding region as the figure). Immediately after observers indicated which interpretation they perceived, a target bar was flashed very briefly (20 ms) on either the region they perceived as figure or the region they perceived as ground. Observers made a speeded response when they detected the target. Using this method, Wong and Weisstein could hold stimulus information constant and assess performance when both the central fixated region and the un-fixated surround region were perceived as figure and as ground.

In Wong and Weisstein's (1982) original experiment, targets were sharp bars. These high spatial frequency targets were detected better when displayed on regions observers perceived as figure than on regions they perceived as grounds; this was true for both the center and surround regions. These results alone could have been explained if more attention were devoted to figures than to grounds during the experiment (cf., Nelson & Palmer, 2007; Peterson & Gibson, 1993; Peterson, Mojica, Salvagio, & Kimchi, 2017). However, Wong and Weisstein (1983) ruled out an interpretation solely in terms of attention by showing that observers were better at detecting low spatial-frequency targets on regions perceived as grounds than on regions perceived as figures.

Taken together, Wong and Weisstein's (1982, 1983) results are consistent with the hypothesis that once a region is perceived as a figure, detailed (i.e., high spatial frequency) analysis is prioritized and once a region is perceived as a ground, coarse (low spatial frequency) analysis is prioritized (Weisstein &Wong, 1986). This differential prioritization of high and low spatial frequency information may arise from the habitual allocation of attention and/or fixation to figures in a three-dimensional world, and consequently, the availability of high spatial frequency information on figures, and the concomitant blurring of details on more distant ground regions (Hochberg, 1971; cf., Alvarez, 2011; Carrasco et al., 2002; Yeshurun & Carrasco, 1998). These associations in the real world establish priors that are relevant to two-dimensional as well as three-dimensional displays. In other words, based on experience, high spatial frequency content is prioritized on figures and low spatial frequency content is prioritized on grounds.

The history of processing high versus low spatial frequency content on figures and grounds established priors for figure assignment as well. Brown and Weisstein (1988) and Klymenko et al. (1989) showed that regions filled with a relatively high spatial frequency grating were perceived as figures substantially more often than regions filled with a lower spatial frequency grating.

Wong and Weisstein (1982, p. 588) interpreted their results as consistent with the hypothesis that, "Figure perception is characterized by high resolution detail processing while ground perception is characterized by low resolution and insensitivity to phase information." Nevertheless, they did not interpret their results as evidence that past experience affects figure assignment. Instead, they interpreted their results in terms of two "systems" of processing, one for figures and one for grounds. They also claimed that figure-ground processing was a dichotomy like that of sustained versus transient channels, and that grounds are processed faster than figures (cf., Calis & Leeuwenberg, 1981). This line of reasoning was confusing to some including this writer, because figure-ground perception is an outcome of visual processing; there is no ground before there is a figure (Peterson, 2003).

In a related line of work, Weisstein, Maguire, and Brannan (1992) showed that when a spatial frequency grating filling one region was colored a diffuse red, that region was more likely to be perceived as figure. Given that some evidence suggests that diffuse red disables processing in the magnocellular pathway (specialized for fast processing of low spatial frequency information), these findings are consistent with an association between activity in the parvocellular pathway and figural status.

The associations revealed by Weisstein's work are clear. They show that spatial frequency, magno- and parvo-cellular activity, and figure-ground status are *coupled*: figures are coupled with high spatial frequency and parvocellular analysis whereas grounds are coupled with low spatial frequency and magnocellular analysis. In my opinion, as I argued above, this coupling is rooted in past experience. Experience perceiving figures establishes memories for the type of processing associated with figures (e.g., high spatial resolution processing consequent to attending to figures (Yeshurun & Carrasco, 1998)); this type of processing in turn becomes a prior for figure assignment. Understood in this fashion, Weisstein's work on figure assignment was ahead of its time, foreshadowing the Bayesian approach to visual perception.

V. Summary and Conclusion

This chapter began with a review of Weisstein's research showing global influences on detection, which supported her argument that global high-level processes influence local lower-level processes. Next, more contemporary research from my laboratory showing global influences and effects of object memories (a form of past experience) on figure assignment was reviewed. Then, Weisstein's research showing that high spatial frequency analysis is coupled with figural status and low spatial frequency analysis is coupled with ground status was reviewed. Weisstein's work on figure assignment can be understood in contemporary, Bayesian, terms as an instance of past experience effects on figure assignment.

Naomi Weisstein combined careful psychophysical research with groundbreaking theory. She was already ill when I was in graduate school. I knew her by reputation, and by her published articles, but I never met her. Her work was innovative then and is still thought-provoking today.

Notes

1 To see color versions of Figures 10.2, 10.3, & 10.6 go to https://research.franklin.uga.edu/visual-perception-laboratory/chapter-10
2 Of course, the B/W fill in the convex and concave regions was balanced across the displays tested.
3 Note that in all the display types tested, convex and concave regions differed in luminance: one was higher in luminance than the medium gray backdrop; the other was lower in luminance. There were >64 displays of each type. For all types, the relative luminance of the convex and concave regions was balanced across the stimuli. Different groups of participants viewed the different types of displays.

References

Alvarez, G. A. (2011). Representing multiple objects as an ensemble enhances visual cognition. *Trends in Cogntive Sciences*, *15* (3), 122–131.

Barense, M. G., Ngo, J., Hung, L., & Peterson, M. A. (2012). Interactions of memory and perception in amnesia: The figure-ground perspective. *Cerebral Cortex*, *22* (11), 2680–2691.

Brown, J. M., & Weisstein, N. (1988). A spatial frequency effect on perceived depth. *Attention, Perception & Psychophysics*, *44* (2), 157–166.

Burge, J., Fowlkes, C. C., & Banks, M. S. (2010). Natural-scene statistics predict how the figure–ground cue of convexity affects human depth perception. *Journal of Neuroscience*, *30* (21), 7269–7280.

Burrola, M., & Peterson, M. A. (2014, May). Global influences on figure assignment: The role of the border. Paper presented at Vision Sciences Society Meeting, St. Pete Beach, FL.

Cacciamani, L., Wager, E., Peterson, M. A., & Scalf, P. E. (2017). Age-related changes in the sensitivity of the perirhinal cortex to configuration and part familiarity and its connectivity to visual cortex. *Frontiers in Aging Neuroscience*, *9*, 291.

Calis, G., & Leeuwenberg, E. (1981). Grounding the figure. *Journal of Experimental Psychology: Human Perception and Performance*, 7 (6), 1386.

Carrasco, M., Williams, P. E., & Yeshurun, Y. (2002). Covert attention increases spatial resolution with or without masks: Support for signal enhancement. *Journal of Vision, 2*, 467–479.

Felleman, D. J., & Van Essen, D. C. (1991). Distributed hierarchical processing in the primate cerebral cortex. *Cerebral Cortex, 1*, 1–47.

Fine, I., MacLeod, D. I., & Boynton, G. M. (2003). Surface segmentation based on the luminance and color statistics of natural scenes, *Journal of the Optical Society of America, A: Optics, Image Science & Vision, 20*, 1283–1291.

Gibson, B. S., & Peterson, M. A. (1994). Does orientation-independent object recognition precede orientation-dependent recognition? Evidence from a cueing paradigm. *Journal of Experimental Psychology: Human Perception and Performance, 20*, 299–316.

Goldreich, D., & Peterson, M. A. (2012). A Bayesian observer replicates convexity context effects. *Seeing and Perceiving, 25* (3–4), 365–395.

Hochberg, J. (1971). Perception. In J. W. Kling & L. A. Riggs (Eds) *Experimental Psychology*, (pp. 395–550). New York: Holt, Rinehart & Winston.

Ing, A. D., Wilson, J. A., & Geisler, W. S. (2010). Region grouping in natural foliage scenes: Image statistics and human performance, *Journal of Vision, 10* (4), 11–19.

Kanisza, G., & Gerbino, W. (1976). Convexity and symmetry in figure–ground organization. In M. Henle (Ed.) *Vision and Artifact*, (pp. 25–32). New York: Springer.

Klymenko, V., & Weisstein, N. (1986). Spatial frequency differences can determine figure-ground organization. *Journal of Experimental Psychology: Human Perception and Performance*, 12 (3), 324–330.

Klymenko, V., Weisstein, N., Topolski, R., & Hsieh, C. H. (1989). Spatial and temporal frequency in figure-ground organization. *Perception & Psychophysics, 45*(5), 395–403.

Lass, J. W., Bennett, P. J., Peterson, M. A., & Sekuler, A. B. (2017). Effects of aging on figure-ground perception: Convexity context effects and competition resolution. *Journal of Vision, 17* (2), 1–16.

Mojica, A. J., & Peterson, M. A. (2014). Display-wide influences on figure-ground perception: The case of symmetry. *Attention, Perception & Psychophysics, 76* (4), 1069–1084.

Nelson, R. A., & Palmer, S. E. (2007). Familiar shapes attract attention in figure–ground displays. *Perception & Psychophysics, 69*, 382–392.

Peterson, M. A. (1994). The proper placement of uniform connectedness. *Psychonomic Bulletin and Review, 1*, 509–514.

Peterson, M. A. (2003). On figures, grounds, and varieties of amodal surface completion. In R. Kimchi, M. Behrmann & C. Olson (Eds) *Perceptual Organization in Vision: Behavioral and Neural Perspectives*, (pp. 87–116). Mahwah, NJ: LEA.

Peterson, M. A., & Gibson, B. S. (1991). The initial identification of figure-ground relationships: Contributions from shape recognition routines. *Bulletin of the Psychonomic Society, 29*, 199–202.

Peterson, M. A., & Gibson, B. S. (1993). Shape recognition contributions to figure-ground organization in three-dimensional displays. *Cognitive Psychology, 25*, 383–429.

Peterson, M. A., & Gibson, B. S. (1994a). Must figure-ground organization precede object recognition? An assumption in peril. *Psychological Science, 5*, 253–259.

Peterson, M. A., & Gibson, B. S. (1994b). Object recognition contributions to figure-ground organization: Operations on outlines and subjective contours. *Perception & Psychophysics, 56*, 551–564.

Peterson, M. A., & Salvagio, E. (2008). Inhibitory competition in figure-ground perception: Context and convexity. *Journal of Vision, 8* (16), 1–13.

Peterson, M. A., & Salvagio, E. (2017). Gestalt demonstrations don't tell the whole story of figure-ground perception. Part II: Convexity. Talk presented at the 58th Annual Meeting of the Psychonomic Society, Vancouver, BC.

Peterson, M. A., Harvey, E. H., & Weidenbacher, H. L. (1991). Shape recognition inputs to figure-ground organization: Which route counts? *Journal of Experimental Psychology: Human Perception and Performance*, *17*, 1075–1089.

Peterson, M. A., Cacciamani, L., Barense, M. D., & Scalf, P. E. (2012). The perirhinal cortex modulates V2 activity in response to the agreement between part familiarity and configuration familiarity. *Hippocampus*, *22*, 1965–1977.

Peterson, M. A., Gerhardstein, P. C., Mennemeier, M., & Rapcsak, S. Z. (1998). Object-centered attentional biases and object recognition contributions to scene segmentation in left- and right-hemisphere-damaged patients. *Psychobiology*, *26*, 557–570.

Peterson, M. A., Mojica, A. J., Salvagio, E., & Kimchi, R. (2017). Figural properties are prioritized for search under conditions of uncertainty: Setting boundary conditions on claims that figures automatically attract attention. *Attention, Perception & Psychophysics*, *79* (1), 180–199.

Peterson, M. A., de Gelder, B., Rapcsak, S. Z., Gerhardstein, P. C., & Bachoud-Lévi, A. C. (2000). Object memory effects on figure assignment: Conscious object recognition is not necessary or sufficient. *Vision Research*, *40*, 1549–1567.

Reicher, G. M. (1969). Perceptual recognition as a function of meaningfulness of stimulus material. *Journal of Experimental Psychology*, 81 (2), 275–280.

Spanò, G., Peterson, M. A., Nadel, L., & Edgin, J. O. (2016). Seeing can be remembering: Interactions between memory and perception in typical and atypical development. *Clinical Psychological Science*, *4* (2), 254–271.

Su, Y. R., He, Z. J., & Ooi, T. L. (2010). Boundary contour-based surface integration affected by color. *Vision Research*, *50* (18), 1833–1844.

Tynan, P., & Sekuler, R. (1975). Moving visual phantoms: A new contour completion effect. *Science*, *188* (4191), 951–952.

Weisstein, N. (1970). Neural symbolic activity: A psychophysical measure. *Science*, *168* (3938), 1489–1491.

Weisstein, N. (1973). Beyond the yellow-Volkswagen detector and the grandmother cell: A general strategy for the exploration of operations in human pattern recognition. In R. L. Solso, (Ed.) *Contemporary Issues in Cognitive Psychology: The Loyola Symposium*. Washington, DC: Winston/Wiley.

Weisstein, N., & Harris, C. S. (1974). Visual detection of line segments: An object-superiority effect. *Science*, *186*, 752–755.

Weisstein, N., & Maguire, W. (1978). Computing the next step: Psychophysical measures of representation and interpretation. In A. R. Hanson & E. M. Riseman (Eds) *Computer Vision Systems*, (pp. 243–260). New York: Academic Press.

Weisstein, N., & Wong, E. (1986). Figure-ground organization and the spatial and temporal responses of the visual system. In E. C. Schwab & H. C. Nusbaum (Eds), *Pattern Recognition by Humans and Machines: Visual Perception Vol. 2* (pp. 31–64). Orlando, FL: Academic Press.

Weisstein, N., Maguire, W., & Berbaum, K. (1977). A phantom-motion aftereffect. *Science*, *198* (4320), 955–958.

Weisstein, N., Maguire, W., & Brannan, J. R. (1992). M and P pathways and the perception of figure and ground. In J. R. Brannan (Ed.), *Applications of Parallel Processing in Vision* (pp. 137–166). New York: Elsevier Science Publishers B.V.

Weisstein, N., Matthews, M. L., & Berbaum, K. (1974, January). Illusory contours can mask real contours. *Bulletin of the Psychonomic Society, 4*, 266.

Weisstein, N., Montalvo, F., & Ozog, G. (1972). Differential adaptation to gratings blocked by cubes and gratings blocked by hexagons: A test of the neural symbolic activity hypothesis. *Psychonomic Science, 27* (2), 89–91.

Wheeler, D. D. (1970). Processes in word recognition. Cognitive Psychology, 1 (1), 59–85.

Wong, E., & Weisstein, N. (1982). A new perceptual context-superiority effect: Line segments are more visible against a figure than against a ground. *Science, 218* (4572), 587–589.

Wong, E., & Weisstein, N. (1983). Sharp targets are detected better against a figure, and blurred targets are detected better against a background. *Journal of Experimental Psychology: Human Perception and Performance, 9* (2), 194.

Wyatte, D., Curran, T., & O'Reilly, R. (2012). The limits of feedforward vision: Recurrent processing promotes robust object recognition when objects are degraded. *Journal of Cognitive Neuroscience, 24* (11), 2248–2261.

Yeshurun, Y., & Carrasco, M. (1998). Attention improves or impairs visual performance by enhancing spatial resolution. *Nature*, 396, 72–75.

Yin, C., Kellman, P. J., & Shipley, T. F. (1997). Surface completion complements boundary interpolation in the visual integration of partly occluded objects. *Perception, 26* (11), 1459–1479.

Yin, C., Kellman, P. J., & Shipley, T. F. (2000). Surface integration influences depth discrimination. *Vision Research, 40* (15), 1969–1978.

11

THE ROLE OF GROUND FEATURES IN THE PERCEPTION OF FIGURE-GROUND AND SUBJECTIVE CONTOURS

Barbara Gillam
University of New South Wales

Figure-ground was one of the many interests of Naomi Weisstein (e.g., Brown & Weisstein, 1989; Klymenko et al., 1989; Klymenko & Weisstein, 1986; Weisstein et al., 1992). Wong and Weisstein (1982), for example, showed that tilt discrimination of a line segment was better when the line was placed on the figure region than on the ground region. In honor of Naomi's interest, the present paper explores some novel aspects of figure-ground perception with an emphasis on the perception of ground. Most of the illustrative examples are taken from Australian Aboriginal painting.

The emphasis in figure-ground research has always been on properties of figure, with ground treated as a default. This begins with Rubin (1915), who showed that areas that are more symmetric, concave and narrow tend to be seen as figure, implying that areas that are asymmetric, convex and wider will be seen as ground. Figure 11.1 is an Australian aboriginal painting showing that the strips that are parallel/narrow appear as figures.[1] They appear to occlude the wider non-parallel brown areas which link up to form a continuous ground.

Contrast

Orientation/Curvature Contrast

Another important factor in figure-ground resolution (but not considered by Rubin) is contrast. In Figure 11.2 the left edge of the black shape on the left differs in curvature/orientation from the straight elements it cuts off and it appears to occlude those elements. On the other hand, the edge of the snake on the right is not independent of surrounding elements and tends to be seen as a hole rather

FIGURE 11.1 Mick Namarari Tjapaltjarri, *Untitled*, 1998.

FIGURE 11.2 Gaiyabidja Lalara, *Untitled*, 1970s.

than an object, despite its snake-like appearance. Meaning per se is not the deciding factor.

Orientation contrast has been investigated experimentally outside the context of art (Gillam, 1987; Gillam & Chan, 2002; Gillam et al., 2014) and shown to be an important factor in producing an impression of occlusion.

Although the figure-ground literature has emphasized the stimulus properties that give rise to figure, there are certain properties of an area that tend to favor perceiving it as ground. These include *entropy contrast* and *extrinsic edge*.

Entropy Contrast

A set of "inducers" (either trapezia or thin lines) whose terminations form a linear alignment may appear occluded at the alignment (see Figure 11.3).

It can be seen that perceived occlusion is weak or non-existent if the inducers are regular (Figure 11.3a). The sense that they are occluded becomes stronger when the inducers differ in width, separation and height (Figure 11.3b) and even stronger when they also differ in orientation (Figure 11.3c). These observations have been confirmed experimentally using both ratings of the strength of

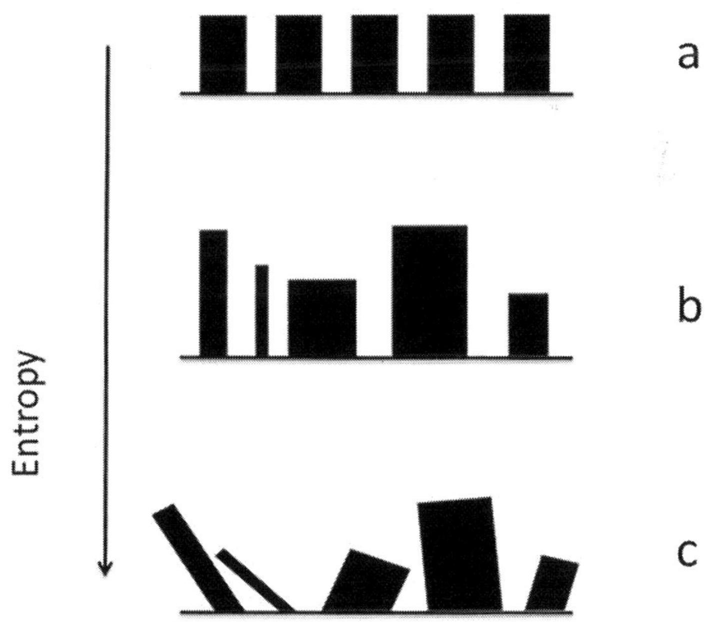

FIGURE 11.3 Entropy (disorder) contrast.

The more diverse the aligned objects the greater the sense of an occlusion at the alignment. [After Gillam and Chan (2002). Reprinted by permission of SAGE Publications, Inc.]

perceived occlusion with a line present along the alignment and the strength of a subjective contour in the absence of a line (Gillam & Chan, 2002; Gillam & Marlow, 2014). The role of inducer irregularity in fostering strong perceived occlusion/subjective contours at their alignment can be understood on ecological grounds. It is highly unlikely that objects or lines with very diverse properties (high entropy) will have aligned terminations except when occluded by a nearer surface. The regularity of their alignment in that case contrasts with the diversity of the individual elements in other respects, hence the term "entropy contrast".

Figure 11.4 shows that even when there is no entropy (all the aligned elements are the same), the sense that they are occluded by a line placed along their alignment is greater when the latter is neither orthogonal to nor parallel to other lines or edges of the element set (Gillam, 2014) i.e., "extrinsic ". The facilitating effect of aligned extrinsic edges is also ecologically based in that they tend to occur when an occluder intersects regular shapes.

Figure 11.5 shows an Australian Aboriginal painting of *The Story of the Wagilag Sisters* by Mawalan Marika (1948). The snakes appear to strongly occlude their background by entropy contrast (as well as curvature contrast) while many of the background strips have extrinsic edges (edges neither orthogonal to nor parallel to the elements they cut off) and appear occluded at these edges.

Delineated Ground

Figure 11.6 shows another Australian Aboriginal painting of figures on ground with the ground areas filled with texture made of lines approximately parallel to each ground outline. Figure 11.7 shows the same phenomenon. Although in natural figure-ground conditions, texture would be independent of ground shape, in these figures the sense of ground persists, probably because the circular figures are connected and the ground areas are not. Delineation of the shape of ground areas is common in Aboriginal painting. This may reflect the importance of noticing

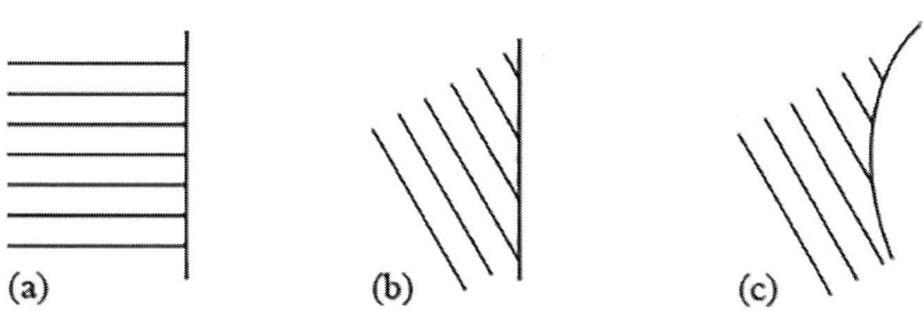

FIGURE 11.4 Examples of (b) orientation and (c) curvature contrast.

The contrasting edges are seen as extrinsic to (not belonging to) the parallel lines. There is a stronger sense of occlusion by the terminating line in (b) and (c) than in (a), in which the edge is orthogonal to the lines.

FIGURE 11.5 Mawalan Marika, *The Wagelag Sisters*, 1948.

 See text for comments.

FIGURE 11.6 Warlimpirrnga Tjapaltjarri, *Dingo Dreaming*, 1988.

 The lines on the ground areas follow the shapes of the figures and their connections.

ground features in Aboriginal life. In Figure 11.7 contrast is also lower for the ground areas, which reduces their salience.

Disordered Ground

Figure 11.8 shows two Aboriginal paintings in which the ground is deliberately disordered, making it difficult to see the figures in parallel; they must be detected

FIGURE 11.7 Pinta Pinta Tjapanangka, *Tingari Dreaming at Yaru Yaru*, c 1978.

© Estate of the artist, licensed by Aboriginal Artists Agency Ltd. The brown and orange ground areas have delineations parallel to the edges of the figures.

FIGURE 11.8 Mawalan Marika, *Tribesmen at Sea and Land*, 1958.

© Estate of the artist, licensed by Aboriginal Artists Agency Ltd. Deliberately disordered ground makes figures difficult to discern.

by serial scrutiny. Again, it is a feature of Aboriginal art to disguise features of spiritual importance, which may be the reason for disordering the ground thus making the figures more difficult to see.

Nested Grounds

In Aboriginal painting there is no linear perspective, but strong perceived depth is achieved by nesting successive grounds, with each occluding surface being itself occluded. Figure 11.9 shows the depth effect produced by such nested occlusions. By combining nesting of the red circular bulls eye shapes with size gradients. Figure 11.10 gives an impression of *movement* in depth.

Pseudo-Transparency

Figure 11.11 shows an interesting effect of transparency produced by similar texture inside and outside the outlines of the fish figures, even though there is no actual continuity of specific texture across the border, which would occur if the fish shapes were indeed transparent. Hence, we call this effect "pseudo-transparency". There is some tension in these figures between the impression of solidity and transparency. Aboriginal art often seems to explore the boundary conditions for perceptual effects.

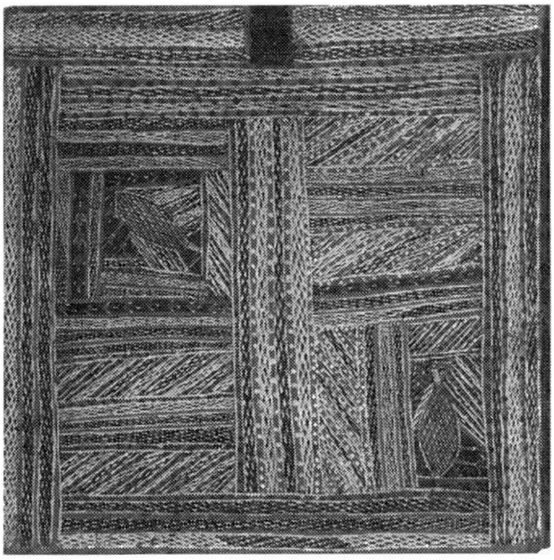

FIGURE 11.9 Mawalan Marika, *Yalangbara*, 1960.

© Estate of the artist, licensed by Aboriginal Artists Agency Ltd. Nested occlusions convey a sense of distance such as for the birds on the upper left and lower right.

FIGURE 11.10 Shorty Lungkata Tungurrayi, *Men Travelling in a Bush Fire*, 1972–1973.

© Estate of the artist, licensed by Aboriginal Artists Agency Ltd. Nested occlusions of red "bulls eye" figures combined with a size gradient give rise to a sense of their motion in depth.

FIGURE 11.11 Marabinyin, *Mullet Fish Totems*.

© Estate of the artist, licensed by Aboriginal Artists Agency Ltd.

Ground Not Independent of Figure

Figure 11.12 shows an emu in its environment. The ground in this case is not independent of the figure but forms a context, which is influenced by the figure.

Changing Ground

In Figure 11.13 some of the figures are seen partially on one ground and partially on another with no loss of figure integrity or grouping.

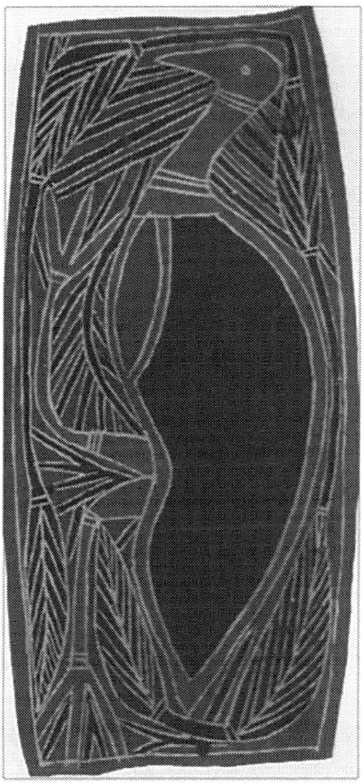

FIGURE 11.12 Peter Mondjingu, *Emu*, c 1970s.

FIGURE 11.13 Yirawala, *Mosquito Dance*, 1976.

Summary

This paper illustrates many ways in which figure and ground relationships can be depicted in art – in this case Australian Aboriginal bark painting. These include the overlapping of one surface by another, nested overlapping, pseudo-transparency, figure influencing ground, the infilling of ground, changing ground, and the possibility of eliminating the impression of figure and ground by introducing surface disorder.

Note

1 To see color versions of all Australian aboriginal paintings for Chapter 11 go to https://research.franklin.uga.edu/visual-perception-laboratory/chapter-11

References

Brown, J. M., & Weisstein, N. (1989). Conflicting figure-ground and depth information reduces moving phantom visibility. *Perception, 20* (2), 155–165.

Gillam, B. (1987). Perceptual grouping and subjective contours. In S. Petry & G. Meyer (Eds) *The Perception of Illusory Contours* (pp. 268–273). New York: Springer.

Gillam, B. (2014). *Grouping and occlusion in perception and art.* The 22nd Kanizsa lecture. Trieste: Trieste University Press.

Gillam, B., & Chan, W. M. (2002). Grouping has a negative effect on both subjective contours and perceived occlusion at T-junctions. *Psychological Science, 13* (3), 279–283.

Gillam, B., & Marlow, P. J. (2014). Comparing subjective contours for Kanizsa squares and linear edge alignments ("New York Titanic" figures). *Perception, 43*, 977–988.

Gillam, B., Wardle, S., & Vecellio, E. (2014). Orientation contrast and entropy contrast in the genesis of subjective contours along thin lines. *Perception, 43* (1), 7–22.

Klymenko, V., & Weisstein, N. (1986). Spatial frequency differences can determine figure-ground organization. *Journal of Experimental Psychology: Human Perception and Performance, 12* (3), 324–330.

Klymenko, V., Weisstein, N., Topolski, R., & Hsieh, C. H. (1989). Spatial and temporal frequency in figure-ground organization. *Perception, 18* (5), 639–648.

Rubin, E. (1915). *Synsoplevede Figurer.* Copenhagen: Gyldendalske Boghandel Nordisk Forlag.

Weisstein, N., Maguire, W., & Brannan, J. R. (1992). M and P pathways and the perception of figure and ground. In J. R. Brannan (Ed.), *Applications of Parallel Processing in Vision* (pp. 137–166). New York: Elsevier Science Publishers B.V.

Wong, E., & Weisstein, N. (1982). A new perceptual context-superiority effect: Line segments are more visible against a figure than against a ground. *Science, 218* (4572), 587–589.

INDEX

Page numbers in italics refer to figures. Page references in bold refer to tables. Page references followed by "n" refer to notes.

Made in the USA
Las Vegas, NV
24 October 2026

40cbfe34-b829-4267-90fe-b81e59d11406R01